Pitman Research Notes in Mathematics Series

Submission of proposals for consideration
Suggestions for publication, in the form of outlines and representative samples, are invited by the Editorial Board for assessment. Intending authors should approach one of the main editors or another member of the Editorial Board, citing the relevant AMS subject classifications. Alternatively, outlines may be sent directly to the publisher's offices. Refereeing is by members of the board and other mathematical authorities in the topic concerned, throughout the world.

Preparation of accepted manuscripts
On acceptance of a proposal, the publisher will supply full instructions for the preparation of manuscripts in a form suitable for direct photo-lithographic reproduction. Specially printed grid sheets are provided and a contribution is offered by the publisher towards the cost of typing. Word processor output, subject to the publisher's approval, is also acceptable.

Illustrations should be prepared by the authors, ready for direct reproduction without further improvement. The use of hand-drawn symbols should be avoided wherever possible, in order to maintain maximum clarity of the text.

The publisher will be pleased to give any guidance necessary during the preparation of a typescript, and will be happy to answer any queries.

Important note
In order to avoid later retyping, intending authors are strongly urged not to begin final preparation of a typescript before receiving the publisher's guidelines and special paper. In this way it is hoped to preserve the uniform appearance of the series.

Longman Scientific & Technical
Longman House
Burnt Mill
Harlow, Essex, UK
(tel (0279) 26721)

Titles in this series

1 Improperly posed boundary value problems
A Carasso and A P Stone

2 Lie algebras generated by finite dimensional ideals
I N Stewart

3 Bifurcation problems in nonlinear elasticity
R W Dickey

4 Partial differential equations in the complex domain
D L Colton

5 Quasilinear hyperbolic systems and waves
A Jeffrey

6 Solution of boundary value problems by the method of integral operators
D L Colton

7 Taylor expansions and catastrophes
T Poston and I N Stewart

8 Function theoretic methods in differential equations
R P Gilbert and R J Weinacht

9 Differential topology with a view to applications
D R J Chillingworth

10 Characteristic classes of foliations
H V Pittie

11 Stochastic integration and generalized martingales
A U Kussmaul

12 Zeta-functions: An introduction to algebraic geometry
A D Thomas

13 Explicit *a priori* inequalities with applications to boundary value problems
V G Sigillito

14 Nonlinear diffusion
W E Fitzgibbon III and H F Walker

15 Unsolved problems concerning lattice points
J Hammer

16 Edge-colourings of graphs
S Fiorini and R J Wilson

17 Nonlinear analysis and mechanics: Heriot-Watt Symposium Volume I
R J Knops

18 Actions of fine abelian groups
C Kosniowski

19 Closed graph theorems and webbed spaces
M De Wilde

20 Singular perturbation techniques applied to integro-differential equations
H Grabmüller

21 Retarded functional differential equations: A global point of view
S E A Mohammed

22 Multiparameter spectral theory in Hilbert space
B D Sleeman

24 Mathematical modelling techniques
R Aris

25 Singular points of smooth mappings
C G Gibson

26 Nonlinear evolution equations solvable by the spectral transform
F Calogero

27 Nonlinear analysis and mechanics: Heriot-Watt Symposium Volume II
R J Knops

28 Constructive functional analysis
D S Bridges

29 Elongational flows: Aspects of the behaviour of model elasticoviscous fluids
C J S Petrie

30 Nonlinear analysis and mechanics: Heriot-Watt Symposium Volume III
R J Knops

31 Fractional calculus and integral transforms of generalized functions
A C McBride

32 Complex manifold techniques in theoretical physics
D E Lerner and P D Sommers

33 Hilbert's third problem: scissors congruence
C-H Sah

34 Graph theory and combinatorics
R J Wilson

35 The Tricomi equation with applications to the theory of plane transonic flow
A R Manwell

36 Abstract differential equations
S D Zaidman

37 Advances in twistor theory
L P Hughston and R S Ward

38 Operator theory and functional analysis
I Erdelyi

39 Nonlinear analysis and mechanics: Heriot-Watt Symposium Volume IV
R J Knops

40 Singular systems of differential equations
S L Campbell

41 N-dimensional crystallography
R L E Schwarzenberger

42 Nonlinear partial differential equations in physical problems
D Graffi

43 Shifts and periodicity for right invertible operators
D Przeworska-Rolewicz

44 Rings with chain conditions
A W Chatters and C R Hajarnavis

45 Moduli, deformations and classifications of compact complex manifolds
D Sundararaman

46 Nonlinear problems of analysis in geometry and mechanics
M Atteia, D Bancel and I Gumowski

47 Algorithmic methods in optimal control
W A Gruver and E Sachs

48 Abstract Cauchy problems and functional differential equations
F Kappel and W Schappacher

49 Sequence spaces
W H Ruckle

50 Recent contributions to nonlinear partial differential equations
H Berestycki and H Brezis

51 Subnormal operators
J B Conway

52 Wave propagation in viscoelastic media
F Mainardi

53 Nonlinear partial differential equations and their applications: Collège de France Seminar. Volume I
H Brezis and J L Lions

54 Geometry of Coxeter groups
H Hiller

55 Cusps of Gauss mappings
T Banchoff, T Gaffney and C McCrory

56 An approach to algebraic K-theory
A J Berrick

57 Convex analysis and optimization
J-P Aubin and R B Vintner

58 Convex analysis with applications in the differentiation of convex functions
J R Giles

59 Weak and variational methods for moving boundary problems
C M Elliott and J R Ockendon

60 Nonlinear partial differential equations and their applications: Collège de France Seminar. Volume II
H Brezis and J L Lions

61 Singular systems of differential equations II
S L Campbell

62 Rates of convergence in the central limit theorem
Peter Hall

63 Solution of differential equations by means of one-parameter groups
J M Hill

64 Hankel operators on Hilbert space
S C Power

65 Schrödinger-type operators with continuous spectra
M S P Eastham and H Kalf

66 Recent applications of generalized inverses
S L Campbell

67 Riesz and Fredholm theory in Banach algebra
B A Barnes, G J Murphy, M R F Smyth and T T West

68 Evolution equations and their applications
F Kappel and W Schappacher

69 Generalized solutions of Hamilton-Jacobi equations
P L Lions

70 Nonlinear partial differential equations and their applications: Collège de France Seminar. Volume III
H Brezis and J L Lions

71 Spectral theory and wave operators for the Schrödinger equation
A M Berthier

72 Approximation of Hilbert space operators I
D A Herrero

73 Vector valued Nevanlinna Theory
H J W Ziegler

74 Instability, nonexistence and weighted energy methods in fluid dynamics and related theories
B Straughan

75 Local bifurcation and symmetry
A Vanderbauwhede

76 Clifford analysis
F Brackx, R Delanghe and F Sommen

77 Nonlinear equivalence, reduction of PDEs to ODEs and fast convergent numerical methods
E E Rosinger

78 Free boundary problems, theory and applications. Volume I
A Fasano and M Primicerio

79 Free boundary problems, theory and applications. Volume II
A Fasano and M Primicerio

80 Symplectic geometry
A Crumeyrolle and J Grifone

81 An algorithmic analysis of a communication model with retransmission of flawed messages
D M Lucantoni

82 Geometric games and their applications
W H Ruckle

83 Additive groups of rings
S Feigelstock

84 Nonlinear partial differential equations and their applications: Collège de France Seminar. Volume IV
H Brezis and J L Lions

85 Multiplicative functionals on topological algebras
T Husain

86 Hamilton-Jacobi equations in Hilbert spaces
V Barbu and G Da Prato

87 Harmonic maps with symmetry, harmonic morphisms and deformations of metrics
P Baird

88 Similarity solutions of nonlinear partial differential equations
L Dresner

89 Contributions to nonlinear partial differential equations
C Bardos, A Damlamian, J I Díaz and J Hernández

90 Banach and Hilbert spaces of vector-valued functions
J Burbea and P Masani

91 Control and observation of neutral systems
D Salamon

92 Banach bundles, Banach modules and automorphisms of C*-algebras
M J Dupré and R M Gillette

93 Nonlinear partial differential equations and their applications: Collège de France Seminar. Volume V
H Brezis and J L Lions

94 Computer algebra in applied mathematics: an introduction to MACSYMA
R H Rand

95 Advances in nonlinear waves. Volume I
L Debnath

96 FC-groups
M J Tomkinson

97 Topics in relaxation and ellipsoidal methods
M Akgül

98 Analogue of the group algebra for topological semigroups
H Dzinotyiweyi

99 Stochastic functional differential equations
S E A Mohammed

100 Optimal control of variational inequalities
V Barbu
101 Partial differential equations and dynamical systems
W E Fitzgibbon III
102 Approximation of Hilbert space operators. Volume II
C Apostol, L A Fialkow, D A Herrero and D Voiculescu
103 Nondiscrete induction and iterative processes
V Ptak and F-A Potra
104 Analytic functions – growth aspects
O P Juneja and G P Kapoor
105 Theory of Tikhonov regularization for Fredholm equations of the first kind
C W Groetsch
106 Nonlinear partial differential equations and free boundaries. Volume I
J I Díaz
107 Tight and taut immersions of manifolds
T E Cecil and P J Ryan
108 A layering method for viscous, incompressible L_p flows occupying R^n
A Douglis and E B Fabes
109 Nonlinear partial differential equations and their applications: Collège de France Seminar. Volume VI
H Brezis and J L Lions
110 Finite generalized quadrangles
S E Payne and J A Thas
111 Advances in nonlinear waves. Volume II
L Debnath
112 Topics in several complex variables
E Ramírez de Arellano and D Sundararaman
113 Differential equations, flow invariance and applications
N H Pavel
114 Geometrical combinatorics
F C Holroyd and R J Wilson
115 Generators of strongly continuous semigroups
J A van Casteren
116 Growth of algebras and Gelfand–Kirillov dimension
G R Krause and T H Lenagan
117 Theory of bases and cones
P K Kamthan and M Gupta
118 Linear groups and permutations
A R Camina and E A Whelan
119 General Wiener–Hopf factorization methods
F-O Speck
120 Free boundary problems: applications and theory, Volume III
A Bossavit, A Damlamian and M Fremond
121 Free boundary problems: applications and theory, Volume IV
A Bossavit, A Damlamian and M Fremond
122 Nonlinear partial differential equations and their applications: Collège de France Seminar. Volume VII
H Brezis and J L Lions
123 Geometric methods in operator algebras
H Araki and E G Effros
124 Infinite dimensional analysis–stochastic processes
S Albeverio
125 Ennio de Giorgi Colloquium
P Krée
126 Almost-periodic functions in abstract spaces
S Zaidman
127 Nonlinear variational problems
A Marino, L Modica, S Spagnolo and M Degiovanni
128 Second-order systems of partial differential equations in the plane
L K Hua, W Lin and C-Q Wu
129 Asymptotics of high-order ordinary differential equations
R B Paris and A D Wood
130 Stochastic differential equations
R Wu
131 Differential geometry
L A Cordero
132 Nonlinear differential equations
J K Hale and P Martinez-Amores
133 Approximation theory and applications
S P Singh
134 Near-rings and their links with groups
J D P Meldrum
135 Estimating eigenvalues with *a posteriori/a priori* inequalities
J R Kuttler and V G Sigillito
136 Regular semigroups as extensions
F J Pastijn and M Petrich
137 Representations of rank one Lie groups
D H Collingwood
138 Fractional calculus
G F Roach and A C McBride
139 Hamilton's principle in continuum mechanics
A Bedford
140 Numerical analysis
D F Griffiths and G A Watson
141 Semigroups, theory and applications. Volume I
H Brezis, M G Crandall and F Kappel
142 Distribution theorems of L-functions
D Joyner
143 Recent developments in structured continua
D De Kee and P Kaloni
144 Functional analysis and two-point differential operators
J Locker
145 Numerical methods for partial differential equations
S I Hariharan and T H Moulden
146 Completely bounded maps and dilations
V I Paulsen
147 Harmonic analysis on the Heisenberg nilpotent Lie group
W Schempp
148 Contributions to modern calculus of variations
L Cesari
149 Nonlinear parabolic equations: qualitative properties of solutions
L Boccardo and A Tesei
150 From local times to global geometry, control and physics
K D Elworthy

151 A stochastic maximum principle for optimal control of diffusions
U G Haussmann
152 Semigroups, theory and applications. Volume II
H Brezis, M G Crandall and F Kappel
153 A general theory of integration in function spaces
P Muldowney
154 Oakland Conference on partial differential equations and applied mathematics
L R Bragg and J W Dettman
155 Contributions to nonlinear partial differential equations. Volume II
J I Díaz and P L Lions
156 Semigroups of linear operators: an introduction
A C McBride
157 Ordinary and partial differential equations
B D Sleeman and R J Jarvis
158 Hyperbolic equations
F Colombini and M K V Murthy
159 Linear topologies on a ring: an overview
J S Golan
160 Dynamical systems and bifurcation theory
M I Camacho, M J Pacifico and F Takens
161 Branched coverings and algebraic functions
M Namba
162 Perturbation bounds for matrix eigenvalues
R Bhatia
163 Defect minimization in operator equations: theory and applications
R Reemtsen
164 Multidimensional Brownian excursions and potential theory
K Burdzy
165 Viscosity solutions and optimal control
R J Elliott
166 Nonlinear partial differential equations and their applications. Collège de France Seminar. Volume VIII
H Brezis and J L Lions
167 Theory and applications of inverse problems
H Haario
168 Energy stability and convection
G P Galdi and B Straughan
169 Additive groups of rings. Volume II
S Feigelstock
170 Numerical analysis 1987
D F Griffiths and G A Watson
171 Surveys of some recent results in operator theory. Volume I
J B Conway and B B Morrel
172 Amenable Banach algebras
J-P Pier
173 Pseudo-orbits of contact forms
A Bahri
174 Poisson algebras and Poisson manifolds
K H Bhaskara and K Viswanath
175 Maximum principles and eigenvalue problems in partial differential equations
P W Schaefer
176 Mathematical analysis of nonlinear, dynamic processes
K U Grusa
177 Cordes' two-parameter spectral representation theory
D F McGhee and R H Picard
178 Equivariant K-theory for proper actions
N C Phillips
179 Elliptic operators, topology and asymptotic methods
J Roe
180 Nonlinear evolution equations
J K Engelbrecht, V E Fridman and E N Pelinovski
181 Nonlinear partial differential equations and their applications. Collège de France Seminar. Volume IX
H Brezis and J L Lions
182 Critical points at infinity in some variational problems
A Bahri
183 Recent developments in hyperbolic equations
L Cattabriga, F Colombini, M K V Murthy and S Spagnolo
184 Optimization and identification of systems governed by evolution equations on Banach space
N U Ahmed
185 Free boundary problems: theory and applications. Volume I
K H Hoffmann and J Sprekels
186 Free boundary problems: theory and applications. Volume II
K H Hoffmann and J Sprekels
187 An introduction to intersection homology theory
F Kirwan
188 Derivatives, nuclei and dimensions on the frame of torsion theories
J S Golan and H Simmons
189 Theory of reproducing kernels and its applications
S Saitoh
190 Volterra integrodifferential equations in Banach spaces and applications
G Da Prato and M Iannelli
191 Nest algebras
K R Davidson
192 Surveys of some recent results in operator theory. Volume II
J B Conway and B B Morrel
193 Nonlinear variational problems. Volume II
A Marino and M K Murthy
194 Stochastic processes with multidimensional parameter
M E Dozzi
195 Prestressed bodies
D Iesan
196 Hilbert space approach to some classical transforms
R H Picard
197 Stochastic calculus in application
J R Norris
198 Radical theory
B J Gardner
199 The C* – algebras of a class of solvable Lie groups
X Wang

200 Stochastic analysis, path integration and dynamics
D Elworthy
201 Riemannian geometry and holonomy groups
S Salamon
202 Strong asymptotics for extremal errors and polynomials associated with Erdös type weights
D S Lubinsky
203 Optimal control of diffusion processes
V S Borkar
204 Rings, modules and radicals
B J Gardner
205 Numerical studies for nonlinear Schrödinger equations
B M Herbst and J A C Weideman
206 Distributions and analytic functions
R D Carmichael and D Mitrović
207 Semicontinuity, relaxation and integral representation in the calculus of variations
G Buttazzo
208 Recent advances in nonlinear elliptic and parabolic problems
P Bénilan, M Chipot, L Evans and M Pierre
209 Model completions, ring representations and the topology of the Pierce sheaf
A Carson
210 Retarded dynamical systems
G Stepan
211 Function spaces, differential operators and nonlinear analysis
L Paivarinta
212 Analytic function theory of one complex variable
C C Yang, Y Komatu and K Niino
213 Elements of stability of visco-elastic fluids
J Dunwoody
214 Jordan decompositions of generalised vector measures
K D Schmidt
215 A mathematical analysis of bending of plates with transverse shear deformation
C Constanda
216 Ordinary and partial differential equations Volume II
B D Sleeman and R J Jarvis
217 Hilbert modules over function algebras
R G Douglas and V I Paulsen
218 Graph colourings
R Wilson and R Nelson
219 Hardy-type inequalities
A Kufner and B Opic
220 Nonlinear partial differential equations and their applications. College de France Seminar. Volume X
H Brezis and J L Lions

B J Gardner (Editor)
University of Tasmania

Rings, modules and radicals

Proceedings of the Hobart Conference, 1987

Copublished in the United States with
John Wiley & Sons, Inc., New York

Longman Scientific & Technical,
Longman Group UK Limited,
Longman House, Burnt Mill, Harlow
Essex CM20 2JE, England
and Associated Companies throughout the world.

Copublished in the United States with
John Wiley & Sons, Inc., 605 Third Avenue, New York, NY 10158

First published 1989

AMS Subject Classification: (Main) 16A21, 17A65, 16A99
(Subsidiary) 16A76, 16A63, 20K99

ISSN 0269-3674

British Library Cataloguing in Publication Data
Rings, modules and radicals.
1. Algebra. Rings. Radicals
I. Gardner, B.J.
512'.4

ISBN 0-582-03125-7

Library of Congress Cataloging-in-Publication Data
Rings, modules and radicals.
(Pitman research notes in mathematics series; 204)
"Hobart conference, August 1987."
1. Associative rings—Congresses. 2. Modules
(Algebra)—Congresses. 3. Radical theory—Congresses.
I. Gardner, B.J. II. Series.
QA251.5.R56 1989 512'.4 88-8458
ISBN 0-470-21311-6

Printed and bound in Great Britain
by Biddles Ltd, Guildford and King's Lynn

Contents

List of contributors

G.L. Booth	Department of Mathematics, University of Zuluand, Private Bag X1001, 3826 Kwadlangezwa, South Africa.
Cheng Fuchang	Department of Mathematics, Guangxi Teachers' University, Guilin, China.
J. Clark	Department of Mathematics and Statistics, University of Otago, P.O. Box 56, Dunedin, New Zealand.
R. Dimitrić	Department of Mathematics, University of Exeter, North Park Road, Exeter EX4 4QE, U.K.
M.W. Evans	Scotch College, Morrison Street, Hawthorn, Victoria 3122, Australia.
J.S. Golan	Department of Mathematics, University of Haifa, 31999 Haifa, Israel.
N.J. Groenewald	Department of Mathematics, University of Port Elizabeth, P.O. Box 1600, 6000 Port Elizabeth, South Africa.
M. Henriksen	Department of Mathematics, Harvey Mudd College, Claremont, CA 91711, U.S.A.
G. Hofer	Institut für Mathematik, Johannes Kepler Universität Linz, A-4040 Linz, Austria.
P.E. Jambor	Department of Mathematics, University of North Carolina, Wilmington, NC 28406, U.S.A.
S. Kyuno	Department of Mathematics, Tohoku Gakuin University, Tagajo, Miyagi 985, Japan.
B.L. Osofsky	Department of Mathematics, Rutgers - The State University, New Brunswick, NJ 08903, U.S.A.

A.D. Sands — Department of Mathematics and Computer Science, The University, Dundee DD1 4HN, U.K.

S. Veldsman — Department of Mathematics, University of Port Elizabeth, P.O. Box 1600, 6000 Port Elizabeth, South Africa.

R. Wiegandt — Mathematical Institute of the Hungarian Academy of Sciences, P.O. Box 127, H-1364 Budapest, Hungary.

Yi Zhong — Department of Mathematics, Guangxi Teachers' University, Guilin, China.

Zhu Yuan-sen — Department of Mathematics, Hebei Teachers' University, Shi-Jia-Zhuang, Prov. Hebei, China.

Preface

The first international conference dealing with radical theory and related matters was held in Keszthely, Hungary, in 1971. More recently, such conferences have become more frequent, and four others, with varying titles but dealing with essentially the same subject matter, have followed in fairly quick succession: Eger, Hungary, 1982; Krems, Austria, 1984; Hobart, Australia, 1987; Sendai, Japan, 1988. Other similar conferences are already being planned, so it can be seen that radical theory as an algebraic subdiscipline is thriving.

The papers in this volume are based on some of the lectures presented at the Hobart conference. There are both survey papers and original results. As well, there is a collection of open problems contributed during the conference's two Problem Sessions.

I thank all conference participants for their contributions, and I must therefore also record my gratitude to the many institutions and organizations around the world which by their financial assistance to individual conference visitors enabled us to assemble such a distinguished group of participants. I am grateful to the Australian Mathematical Society for a Special Interest Meeting Grant and to the Mathematics Department of the University of Tasmania for some financial support given to invited speakers; for their assistance during the conference and in the preparation of these Proceedings, Betty Golding, Anne Greenhalgh and Mike Parmenter deserve special mentions.

B.J. Gardner
University of Tasmania, Hobart
November 1988

G.L. BOOTH

Jacobson radicals of Γ-near-rings

ABSTRACT: Γ-near-rings were defined by Satyanarayana, and the concept of the left operator near-ring L of a Γ-near-ring M was introduced by the present author. In this paper, we consider zerosymmetric Γ-near-rings only. Three Jacobson-type radicals J_0, J_1 and J_2 are defined. In addition, $J_{1/2}(M)$ is the intersection of the maximal modular left ideals of M. It is shown that $J_0(M) \subseteq J_{1/2}(M) \subseteq J_1(M) \subseteq J_2(M)$. Moreover, $J_2(M) = J_2(L)^+$ and $J_k(L)^+ \subseteq J_k(M)$ for k = 0, 1. If M has a strong left unity, then $J_k(L) = J_k(M)$ for k = 0, 1. ($J_k(L)$ denotes the respective Jacobson-type radical of the near-ring L.)

1. PRELIMINARIES

In this paper, the term "near-ring" will mean a right (distributive) near-ring. A Γ-near-ring is a triple (M, +, Γ) where

(i) (M,+) is a (not necessarily abelian) group;

(ii) Γ is a nonempty set of binary operators on M, such that, for each $\gamma \in \Gamma$, $(M,+,\gamma)$ is a near-ring;

(iii) $x\gamma(y\mu z) = (x\gamma y)\mu z$ for all $x,y,z \in M$ and $\gamma,\mu \in \Gamma$.

Let M and M' be Γ-near-rings. A map $f:M \to M'$ is called a Γ-near-ring homomorphism if, for all $x,y \in M$ and $\gamma \in \Gamma$, $f(x+y) = f(x) + f(y)$ and $f(x\gamma y) = f(x)\gamma f(y)$. The kernels of the Γ-near-ring homomorphisms are precisely those subsets A of M such that

(i) (A,+) is a normal divisor of (M,+);

(ii) for all $a \in A$, $\gamma \in \Gamma$ and $x,y \in M$, $x\gamma(y+a) - x\gamma y \in A$ and $a\gamma x \in A$.

Such a subset A of M is called an ideal of M, denoted $A \triangleleft M$. The same notation will be used for near-rings. If $A \triangleleft M$, the factor group M/A is a Γ-near-ring with the operation $(x+A)\gamma(y+A) = x\gamma y+A$ ($x,y \in M$, $\gamma \in \Gamma$).

Γ-near-rings constitute a variety of Ω-groups as defined by Higgins [2]. This means, *inter alia*, that the usual isomorphism theorems for rings hold for Γ-near-rings ([2], Theorems 3A-3D). Subdirect sums are defined as for rings.

The left operator near-ring L of a Γ-near-ring M was defined in [1]. Let L denote the near-ring of all mappings of M into itself, mappings being taken to operate on the left. If $x \in M$, $\gamma \in \Gamma$, define $[x,\gamma]:M \to M$ by $[x,\gamma]y = x\gamma y$ $(y \in M)$. L is the near-ring generated by the set $\{[x,\gamma]: x \in M, \gamma \in \Gamma\}$.

If $A \subseteq L$, $A^{+} = \{x \in M: [x,\gamma] \in A \text{ for all } \gamma \in \Gamma\}$.

If $B \subseteq M$, $B^{+1} = \{\ell \in L: \ell x \in B \text{ for all } x \in M\}$.

It can be shown ([1], Proposition 3) that these mappings take ideals onto ideals. Moreover, they preserve intersections. If $A \triangleleft L$, then $A \subseteq (A^{+})^{+1}$, and if $B \triangleleft M$, then $B \subseteq (B^{+1})^{+}$.

It can be shown that, if $\ell \in L$, $x \in M$ and $\gamma \in \Gamma$, then $\ell[x,\gamma] = [\ell x,\gamma]$. This identity, which is a consequence of the right distributivity of M, will be of frequent use in the sequel.

A γ-near-ring M is called zerosymmetric if $x\gamma 0 = 0$ for all $x \in M$ and $\gamma \in \Gamma$.

<u>PROPOSITION 1.1</u>: Let M be a Γ-near-ring with left operator near-ring L. Then M is zerosymmetric if and only if L is zerosymmetric.

<u>PROOF</u>: Suppose M is zerosymmetric. Let 0_L be the zero element of L. Then it is easily verified that $0_L = [0,\gamma]$ for any $\gamma \in \Gamma$. Let $\ell \in L$, $x \in M$. Then $\ell[0,\gamma]x = \ell([0\gamma x) = \ell 0$. It follows from the zerosymmetry and right distributivity of M that $\ell 0 = 0$. Hence, $\ell 0_L = 0_L$, as required.

Conversely, suppose that L is zerosymmetric. Let $x \in M$, $\gamma \in \gamma$. Then $x\gamma 0 = x\gamma(0\gamma 0) = [x,\gamma][0,\gamma]0 = [x,\gamma]0_L 0 = 0_L 0 = 0$, since L is zerosymmetric. □

For the rest of this paper, M will denote a zerosymmetric Γ-near-ring with left operator near-ring L. The notation, definitions and terminology relating to near-rings will be as in Pilz [5], unless otherwise stated. For all details pertaining to Γ-rings and their operator rings, we refer to [3].

2. $M\Gamma$-GROUPS

Let $(G,+)$ be a group. If, for all $x,y \in M$, $\gamma,\mu \in \Gamma$ and $g \in G$:

(i) $x\gamma g \in G$;

(ii) $(x+y)\gamma g = x\gamma g + y\gamma g$;

(iii) $x\gamma(y\mu g) = (x\gamma y)\mu g$;

then G is called an $M\Gamma$-group.

REMARK: This concept was first introduced by Satyanarayana [7], who used the term $M\Gamma$-module. We choose the term $M\Gamma$-group for the sake of consistency with Pilz [5].

Let G, G' be $M\Gamma$-groups. If $f:G \to G'$ is a group homomorphism such that, for all $x \in M$, $\gamma \in \Gamma$ and $g \in G$, $f(x\gamma g) = x\gamma f(g)$, then f is called an $M\Gamma$-homomorphism. If f is bijective as well, then it is called an $M\Gamma$-isomorphism. If G is an $M\Gamma$-group, then it may be shown that the kernels of the homomorphisms acting on G are precisely those subsets H of G such that:

(i) $(H,+)$ is a normal divisor of $(G,+)$;

(ii) for all $x \in M$, $\gamma \in \Gamma$, $h \in H$ and $g \in G$, $x\gamma(g+h) - x\gamma g \in H$.

In this case, H is called an $M\Gamma$-ideal of G. A subgroup K of G is called an $M\Gamma$-subgroup of G if, for all $x \in M$, $\gamma \in \Gamma$ and $k \in K$, $x\gamma k \in K$. Since we are considering zerosymmetric Γ-near-rings only, every $M\Gamma$-ideal of G is also an $M\Gamma$-subgroup of G.

Let 0_G denote the zero element of G. Then, if $x \in M$, $\gamma \in \Gamma$ and $g \in G$:

(i) $0\gamma g = 0_G$;

(ii) $x\gamma 0_G = 0_G$.

The second statement is a consequence of the zerosymmetry of M.

If $\gamma \in \Gamma$ and $g \in G$, then $M\gamma g = \{x\gamma g: x \in M\}$. G is said to be monogenic if there exist $g \in G$ and $\gamma \in \Gamma$ such that $M\gamma g = G$. If G is monogenic, and for all $g \in G$, $\gamma \in \Gamma$, either $M\gamma g = 0$ or $M\gamma g = G$, then G is called strongly monogenic.

A monogenic $M\Gamma$-group G is said to be:

(i) of type 0 if G has no $M\Gamma$-ideals except 0 and G itself;

(ii) of type 1 if G is strongly monogenic and has no $M\Gamma$-ideals except 0 and G itself:

(iii) of type 2 if G has no $M\Gamma$-subgroups except 0 and G itself.

It is easily verified that type 2 → type 1 → type 0.

The annihilator of an $M\Gamma$-module G, $\text{ann}_{M\Gamma}G$, is defined to be

$$\text{ann}_{M\Gamma}G = \{x \in M: x\gamma g = 0 \text{ for all } \gamma \in \Gamma, g \in G\}.$$

It is easily verified that $\text{ann}_{M\Gamma}G \triangleleft M$.

A is said to be a k-primitive ideal of M if $A = \text{ann}_{M\Gamma}G$ for some $M\Gamma$-group G of type k ($k = 0, 1, 2$). M is a k-primitive Γ-near-ring if 0 is a k-primitive ideal of M.

We define the Jacobson-type radicals of M as for near-rings, i.e.

$$J_k(M) = \cap\{A: A \text{ is a k-primitive ideal of } M\}.$$

Clearly, $J_0(M) \subseteq J_1(M) \subseteq J_2(M)$.

<u>REMARK</u>: The notion of an irreducible right ΓM-module of a Γ-ring M was defined by Kyuno [3], and independently by Ravisankar and Shukla [6], who showed that the Jacobson radical J(M) of M coincides with the intersection of the annihilators of the irreducible ΓM-modules. Left $M\Gamma$-modules of M are similarly defined. Since the left and right Jacobson radicals of a Γ-ring M coincide [4], J(M) is also equal to the intersection of the annihilators of the irreducible $M\Gamma$-modules of M. It is fairly easy to verify that, for

a Γ-ring M, the notions of $M\Gamma$-group of type k (k = 0, 1, 2) coincide with that of an irreducible $M\Gamma$-module. Hence, in this case $J(M) = J_0(M) = J_1(M) = J_2(M)$, i.e. J_0, J_1 and J_2 all generalize the notion of the Jacobson radical of a Γ-ring.

A strong left unity for M is a pair (d,δ) of $M \times \Gamma$ such that $d\delta x = x$ for all $x \in M$.

PROPOSITION 2.1: Suppose that M has a strong left unity (d,δ). Then:

$$J_1(M) = J_2(M).$$

PROOF: Suppose that G is an $M\Gamma$-group of type 1. Let $g \in G$ and $\gamma \in \Gamma$ be such that $G = M\gamma g$. Then if $h \in G$, $h = x\gamma g$ for some $x \in M$. Hence, $d\delta h = d\delta(x\delta g) = (d\delta x)\gamma g = x\gamma g = h$. Now suppose $0 \neq H$ is an $M\Gamma$-subgroup of G. Let $0 \neq h \in H$. Then $M\delta h \neq 0$ since $h = d\delta h$. Since G is strongly monogenic, $M\delta h = G$. But $M\delta h \subseteq H$, since H is an $M\Gamma$-subgroup of G. Hence, G is an $M\Gamma$-group of type 2. Since every $M\Gamma$-group of type 2 is of type 1, the notions of type 1 and type 2 coincide in this case. The result now follows from the definitions of $J_1(M)$ and $J_2(M)$. □

Let I be a left ideal of M. Then I is called modular if there exist $e \in M$, $\varepsilon \in \Gamma$ such that $x\varepsilon e - x \in I$ for all $x \in M$. It is easily seen that, if I is a maximal modular left ideal of M, then M/I is an $M\Gamma$-group of type 0 with the operation $x\gamma(y+I) = x\gamma y+I$ $(x,y \in M, \gamma \in \Gamma)$, and that $\mathrm{ann}_{M\Gamma} M/I \subseteq I$. We define

$$J_{1/2}(M) = \cap I,$$

where the intersection runs over the maximal modular left ideals of M. Suppose that G is an $M\Gamma$-group of type 0. Suppose that $\varepsilon \in \Gamma$ and $g \in G$ are such that $M\varepsilon g = G$. Then define $f: M \to G$ by $f(x) = x\varepsilon g$ $(x \in M)$. It is easily demonstrated that f is an $M\Gamma$-homomorphism of M (considered as an $M\Gamma$-group) onto G. Let I be the kernel of f. Then, since $M\varepsilon g = G$, there exists $e \in M$ such that $e\varepsilon g = g$. Let $x \in M$. Then $f(x\varepsilon e - x) = (x\varepsilon e - x)\varepsilon g = x\varepsilon e\varepsilon g - x\varepsilon g = x\varepsilon g - x\varepsilon g = 0$. Hence, $x\varepsilon e - x \in I$, i.e. I is a modular ideal of M. By [2], Theorem 3B, M/I is $M\Gamma$-isomorphic to G. From these considerations, it follows that

$$J_0(M) \subseteq J_{1/2}(M).$$

A modular left ideal K of M is called k-modular (k = 1, 2) if the factor group M/K is an MΓ-group of type k. It is easily shown that in this case K is a maximal modular ideal of M.

PROPOSITION 2.2:

$$J_k(M) = \cap K,$$

where K runs over the k-modular left ideals of M (k = 1, 2).

PROOF: Let G be an MΓ-group. Using arguments similar to those employed immediately before the proof of this result, it may be shown that G is of type k (k = 1, 2) if and only if G is MΓ-isomorphic to M/K, where K is a k-modular left ideal of M, and that $\mathrm{ann}_{M\Gamma}G = \mathrm{ann}_{M\Gamma}M/K \subseteq K$. Hence,

$$J_k(M) \subseteq \cap K.$$

Let G be an MΓ-group of type k. Then, if $\gamma \in \Gamma$ and $g \in G$, we define $a(\gamma,g) = \{x \in M: x\gamma g = 0\}$. It is easily shown that $a(\gamma,g)$ is a left ideal of M. Moreover

$$\mathrm{ann}_{M\Gamma}G = \bigcap_{\substack{\gamma\in\Gamma \\ g\in G}} a(\gamma,g).$$

Let $G' = \{(\gamma,g) \in \Gamma \times G: M\gamma g = G\}$. Since G is strongly monogenic, for $(\gamma,g) \in \Gamma \times G$, either $(\gamma,g) \in G'$ or $M\gamma g = 0$. It follows that

$$\mathrm{ann}_{M\Gamma}G = \bigcap_{(\gamma,g)\in G'} a(\gamma,g).$$

Now if $(\gamma,g) \in G'$ we define $f: M \to G$ by $f(x) = x\gamma g$. It follows from the definition of G' that f is an MΓ-homomorphism of M onto G. Moreover, the kernel of f is $a(\gamma,g)$. Hence, $M/a(\gamma,g)$ is MΓ-isomorphic to G. It follows that $a(\gamma,g)$ is a k-modular left ideal of M. Hence,

$$J_k(M) = \bigcap_{G} \bigcap_{(\gamma,g)\in G'} a(\gamma,g),$$

where G runs over the $M\Gamma$-groups of type k. Thus $J_k(M)$ is the intersection of a family of k-modular left ideals of M. It follows that

$$\cap K \subseteq J_k(M).$$

This completes the proof. □

COROLLARY 2.3:

$$J_0(M) \subseteq J_{1/2}(M) \subseteq J_1(M) \subseteq J_2(M).$$

PROOF: The fact that $J_{1/2}(M) \subseteq J_1(M)$ follows immediately from Proposition 2.2. The other inclusions have already been proved. □

3. RELATIONSHIPS WITH THE LEFT OPERATOR NEAR-RING

Let $L_0 = \{[x,\gamma]: x \in M, \gamma \in \Gamma\}$. An L_0-representation in L is an algebraically meaningful expression containing symbols from the set $L_0 \cup \{(,),+,-\}$. The weight of the expression is the number of elements of L_0 contained in it. Clearly, every element ℓ of L has at least one (an possibly many) L_0-representations. If $\ell \in L$ has an L_0-representation of weight n, but none of weight $< n$, then ℓ is said to be of rank n. We note that if $\ell \in L$ and rank $\ell = n \geq 2$, then there exists $\ell_1,\ell_2 \in L$ such that rank $\ell_i < n$ $(i = 1, 2)$ and either $\ell = \ell_1 + \ell_2$ or $\ell = \ell_1\ell_2$.

Now let G be a monogenic $M\Gamma$-group. If $g \in G$, there exist $x \in M$, $\gamma \in \Gamma$ and $h \in G$ such that $g = x\gamma h$. If $\ell \in L$, define $\ell g = (\ell x)\gamma h$. We show that this is a well-defined operation. For suppose that $y \in M$, $\mu \in \Gamma$ and $k \in G$ are such that $g = x\gamma h = y\mu k$. We will show by induction on rank ℓ that

$$(\ell x)\gamma h = (\ell y)\mu k. \tag{1}$$

If rank $\ell = 1$, $\ell = [a,\alpha]$ for some $a \in M$ and $\alpha \in \Gamma$. It is easily verified that (1) holds in this case. Now suppose (1) holds whenever rank $\ell < n$.

Let rank $\ell = n$. Then there exist $\ell_1, \ell_2 \in L$ such that rank $\ell_i < n$ $(i = 1, 2)$ and either $\ell = \ell_1 + \ell_2$ or $\ell = \ell_1\ell_2$. If $\ell = \ell_1 + \ell_2$, then

$$
\begin{aligned}
(\ell x)\gamma h &= ((\ell_1 + \ell_2)x)\gamma h \\
&= (\ell_1 x + \ell_2 x)\gamma h \\
&= (\ell_1 x)\gamma h + (\ell_2 x)\gamma h \\
&= (\ell_1 y)\mu k + (\ell_2 y)\mu k \\
&= (\ell y)\mu k.
\end{aligned}
$$

A similar argument shows that (1) holds in the case $\ell = \ell_1\ell_2$. It is easily shown that G is an L-group with respect to this operation and the addition operation defined on G. Moreover, since G is a monogenic $M\Gamma$-group, $G = M\gamma g$ for some $\gamma \in \Gamma$ and $g \in G$. But $Lg \supseteq M\gamma g$, whence $Lg = G$. Hence, G is a monogenic L-group.

Suppose that G is a strongly monogenic $M\Gamma$-group. Then, as we have seen, G is a monogenic L-group. Let $g \in G$. If $M\gamma g = 0$ for all $\gamma \in \Gamma$, then it may be shown by induction on rank ℓ that $\ell g = 0$ for all $\ell \in L$. (The argument is routine, but makes use of the zerosymmetry of L and hence of Proposition 1.1.) Hence, $Lg = 0$ in this case. If $g \in G$ and $\gamma \in \Gamma$ are such that $M\gamma g = G$, then $M\gamma g \subseteq Lg$, whence $Lg = G$. Hence, G is a strongly monogenic L-group.

Now let H be an L-ideal of G. Then, if $h \in H$, $g \in G$, $x \in M$ and $\gamma \in \Gamma$:

$$x\gamma(g+h) - x\gamma g = [x,\gamma](g+h) - [x,\gamma]g \in H.$$

Hence, H is an $M\Gamma$-ideal of G. Similarly, if K is an L-subgroup of G, then K is an $M\Gamma$-subgroup of G.

PROPOSITION 3.1: Let A be a k-primitive ideal of M $(k = 0, 1, 2)$. Then:

(a) A^{+1} is a k-primitive ideal of L;

(b) $(A^{+1})^{+} = A$.

PROOF: (a) Let $A = \mathrm{ann}_{M\Gamma} G$, where G is an $M\Gamma$-group of type k. By the preceding discussion, G is an L-group of type K. Let $\mathrm{ann}_L G$ denote the annihilator of G in L. Let $\ell \in \mathrm{ann}_L G$. Then if $x \in M$, $\gamma \in \Gamma$ and $g \in G$:

$$(\ell x)\gamma g = \ell[x,\gamma]g = 0.$$

Hence $\ell x \in A$ and thus $\mathrm{ann}_L G \subseteq A^{+1}$.

Now suppose $\ell \in A^{+1}$. Let $g \in G$ and $\gamma \in \Gamma$ be such that $G = M\gamma g$. Let $h \in G$. Then there exists $x \in M$ such that $h = x\gamma g$. Hence $\ell h = \ell(x\gamma g) = (\ell x)\gamma g = 0$ since $\ell x \in A$. Thus $\ell \in \mathrm{ann}_L G$. Consequently, $A^{+1} = \mathrm{ann}_L G$, whence A^{+1} is k-primitive in L, as required.

(b) It is easily verified that

$$(A^{+1})^{+} = \{x \in M: x\gamma y \in A \text{ for all } \gamma \in \Gamma\ ,\ y \in M\}.$$

Since $A \lhd M$, clearly $A \subseteq (A^{+1})^{+}$. Let $x \in (A^{+})^{+1}$. Let $g \in G$, $\nu \in \Gamma$ be such that $M\nu g = G$. Let $\mu \in \Gamma$ and $h \in G$. Then there exists $y \in M$ such that $h = y\nu g$. Hence, $x\mu h = x\mu(y\nu g) = (x\mu y)\nu g = 0$ since $x\mu y \in A$. Hence. $(A^{+1})^{+} \subseteq A$, as required. □

Now suppose that G is an L-group. It is easily seen that G is an $M\Gamma$-group with the operation $x\gamma g = [x,\gamma]g$ $(x \in M,\ \gamma \in \Gamma,\ g \in G)$. Let H be an $M\Gamma$-subgroup of G. Then, for all $x \in M$, $\gamma \in \Gamma$ and $h \in H$, $[x,\gamma]h = x\gamma h \in H$. It follows by induction on rank ℓ that $\ell h \in H$ for all $\ell \in L$, i.e. H is an L-subgroup of G.

PROPOSITION 3.2: Let A be a 2-primitive ideal of L. Then

(a) A^{+} is a 2-primitive ideal of M;

(b) $(A^{+})^{+1} = A$.

PROOF: (a) Let G be an L-group of type 2 such that $A = \mathrm{ann}_L G$. By the preceding argument, G is an $M\Gamma$-group, and has no $M\Gamma$-subgroups except 0 and G. Let $g \in G$, $\gamma \in \Gamma$, $x \in M$ and $\ell \in L$. Then $\ell(x\gamma g) = [x,\gamma]g = [\ell x,\gamma]g = (\ell x)\gamma g$. It follows that $M\gamma g$ is an L-subgroup of G and hence that $M\gamma g = 0$ or G. If

$M\gamma g = 0$ for all $\gamma \in \Gamma$ and $g \in G$, then $Lg = 0$ for all $g \in G$, which contradicts the fact that G is a monogenic L-group. Hence, there exist $\gamma \in \Gamma$ and $g \in G$ such that $M\gamma g = G$, i.e. G is a monogenic $M\Gamma$-group.

Thus, G is an $M\Gamma$-group of type 2. Now

$$\begin{aligned} \text{ann}_{M\Gamma} G &= \{x \in M: x\gamma g = 0 \text{ for all } \gamma \in \Gamma,\ g \in G\} \\ &= \{x \in M: [x,\gamma]g = 0 \text{ for all } \gamma \in \Gamma,\ g \in G\} \\ &= \{x \in M: [x,\gamma] \in A \text{ for all } \gamma \in \Gamma\} \\ &= A^+. \end{aligned}$$

Hence, A^+ is a 2-primitive ideal of M, as required.

(b) It is easily shown that

$$(A^+)^{+1} = \{\ell \in L: \ell[x,\gamma] \in A \text{ for all } x \in M,\ \gamma \in \Gamma\}.$$

Since $A \triangleleft L$, $A \subseteq (A^+)^{+1}$. Suppose that $\gamma \in \Gamma$ and $g \in G$ are such that $G = M\gamma g$. Then, if $h \in G$, $h = x\gamma g$ for some $x \in M$. Let $\ell \in (A^+)^{+1}$. Then $\ell h = \ell(x\gamma g) = \ell[x,\gamma]g = 0$ since $\ell[x,\gamma] \in A$. Hence $(A^{+1})^+ \subseteq A$, and the proof is complete. □

PROPOSITION 3.3:

(a) The mapping $A \to A^+$ defines a one-to-one correspondence between the 2-primitive ideals of L and those of M.

(b) $J_2(L)^+ = J_2(M)$.

PROOF:

(a) is an immediate consequence of Propositions 3.1 and 3.2.

(b)
$$\begin{aligned} J_2(L)^+ &= (\cap \{A: A \text{ is a 2-primitive ideal of } L\})^+ \\ &= \cap \{A: A \text{ is a 2-primitive ideal of } L\}^+ \\ &= \cap \{A: A \text{ is a 2-primitive ideal of } M\} \\ &= J_2(M). \quad □ \end{aligned}$$

PROPOSITION 3.4:

(a) $J_k(L)^+ \subseteq J_k(M)$ $(k = 0, 1)$.

(b) If M has a strong left unity (d,δ) then $J_0(L)^+ = J_0(M)$.

PROOF: (a) Let A be a k-primitive ideal of M. Then by Proposition 3.1(a), A^{+1} is a k-primitive ideal of L. Hence, $J_k(L) \subseteq A^{+1}$. It follows that $J_k(L)^+ \subseteq (A^{+1})^+ = A$ by Proposition 3.2(b). Taking intersections as A runs through the k-primitive ideals of M, $J_k(L)^+ \subseteq J_k(M)$ as required.

(b) Suppose that G is a 0-primitive ideal of L. Let G be an L-group of the type 0 such that $\text{ann}_L G = B$. Let $\ell \in L$. Then $\ell = \ell[d,\delta] = [\ell d,\delta]$. Hence, every element of L is expressible in the form $[x,\delta]$, $x \in M$. Since G is monogenic as an L-group, there exists $g \in G$ such that $Lg = G$. It follows that $M\delta g = G$, i.e. G is monogenic as an $M\Gamma$-group.

Now suppose that H is an $M\Gamma$-ideal of G. Then let $\ell \in L$, $g \in G$ and $h \in H$. Let $\ell = [x,\delta]$, $x \in M$. Then:

$$\ell(g+h) - \ell g = [x,\delta](g+h) - [x,\delta]g$$

$$= x\delta(g+h) - x\delta g \in H.$$

Hence, H is an L-ideal of G. It follows that H = 0 or G, and hence that G is an $M\Gamma$-group of type 0. Using reasoning similar to that used in the proof of Proposition 3.2(a) it may be shown that $\text{ann}_{M\Gamma} G = B^+$. Hence, B^+ is a 0-primitive ideal of M. Consequently, $J_0(M) \subseteq B^+$. Taking intersections over the 0-primitive ideals of L, we have that $J_0(M) \subseteq J_0(L)^+$ and hence that $J_0(L)^+ = J_0(M)$ by (a) above. □

REFERENCES

[1] G.L. Booth, A note on Γ-near-rings, Stud. Sci. Math. Hung., to appear.

[2] P.J. Higgins, Groups with multiple operators, J. Lond. Math. Soc. 3 No. 6 (1956), 366-416.

[3] S. Kyuno, On the semi-simple gamma rings, Tôhuku Math. J. 29 (1977), 217-255.

[4] S. Kyuno, Coincidence of the right Jacobson radical and the left Jacobson radical in a gamma ring, Tsukuba J. Math. 3, No. 1 (1959), 31-35.

[5] G. Pilz, Near-Rings, revised edition, North-Holland, Amsterdam, 1983.

[6] S. Ravisankar and U. Shukla, Structure of Γ-rings, Pacific J. Math. 80, No. 2 (1979), 537-559.

[7] Bh. Satyanarayana, Contributions to near-ring theory, Doctoral Thesis, Nagarjuna University, 1984.

G.L. Booth
Department of Mathematics
University of Zululand
Private Bag X1001
3886 Kwadlangezwa
South Africa

CHENG FUCHANG, XU JINZHONG & YI ZHONG

Semilocal rings and modules

0. INTRODUCTION

Let R be a commutative ring with identity element. R is called a local ring if it has only one maximal ideal; R is called a semilocal ring if it has finitely many maximal ideals. This paper mainly deals with semilocal rings and the homological properties of modules. Some important results of regular local rings are generalized to the semilocal case. Meanwhile, we have discussed the structures of regular semilocal rings with a total dimension t.dim R = 1, 2, 3 and structures of rings with global dimension gl. dim R = 1. Finally, semilocal rings with the AR-property are considered, and some interesting results are obtained by homological methods.

1. MODULES OVER SEMILOCAL RINGS

Let R be a semilocal ring, J the Jacobson radical of R. R is called indecomposable if it has no non-trivial idempotent element (except for 0 and 1).

THEOREM 1.1: Let R be an indecomposable semilocal ring and let P be a finitely generated f.g. projective R-module. Then there exists a positive integer r such that $P \simeq R^r$, namely P is a free module with rank r.

PROOF: Let $m_1, m_2, \ldots, m_n$ be all the maximal ideals of R. We know that P has rank because R is indecomposable [1]. Therefore, P_{m_i} and P_{m_j} have the same rank r. Since R_{m_i} is a local ring with maximal ideal $(m_i)_{m_i}$, P_{m_i} is a f.g. projective module over the commutative ring R_{m_i} with rank r. Hence there is an isomorphism as $R_{m_i}/(m_i)_{m_i}$-modules

$$P_{m_i}/(m_i)_{m_i}P_{m_i} \simeq (R_{m_i}/(m_i)_{m_i})^{(Y)}$$

where $i = 1, 2, \ldots, n$.

On the other hand, R/J is a semisimple artinian ring. Hence P/JP is a f.g. semisimple R/J module. Then there is a decomposition

$$P/JP \simeq (R/m_1)^{r_1} \oplus \ldots \oplus (R/m_n)^{r_n}. \tag{1}$$

By localization, we have $(m_i)_{m_j} = R_{m_j}$ if $i \neq j$ and $(R/m_j)_{m_i} \simeq R_{m_i}/(m_j)_{m_i} = 0$. According to (1), $(P/JP)_{m_i} \simeq (R_{m_i}/(m_i)_{m_i})^{r_i}$. Moreover, we have

$$(J)_{m_i} = (m_1 \cap \ldots \cap m_n)_{m_i} = \bigcap_{j=1}^{n} (m_j)_{m_i} = (m_i)_{m_i}.$$

Therefore,

$$(P/JP)_{m_i} = P_{m_i}/(JP)m_i = P_{m_i}/J_{m_i} P_{m_i}$$

$$= P_{m_i}/(m_i)_{m_i} P_{m_i} \simeq (R_{m_i}/(m_i)_{m_i})^r$$

and

$$(R_{m_i}/(m_i)_{m_i})^{r_i} \simeq (R_{m_i}/(m_i)_{m_i})^r.$$

Subsequently, it follows from this that $r_1 = r_2 = \ldots = r_n$ and

$$P/JP = (R/m_1 \oplus \ldots \oplus R/m_n)^r \simeq (R/J)^r.$$

Finally, considering the natural homomorphism $\pi: P \to P/JP$, $\ker \pi = JP$. It is easy to prove that (P,π) is the projective cover of the R-module P/JP, because JP is superfluous. By the uniqueness of projective cover and the isomorphism $R^{(r)}/J^{(r)} \simeq (R/J)^r$ we have $P \simeq R^r$. □

COROLLARY: Every f.g. projective R-module is free iff R is indecomposable.

PROOF: By Theorem 1.1, it is only required to prove that the condition is necessary.

Assume R to be decomposable, and to have an idempotent element e ($e \neq 0, 1$). It implies that $R = R\cdot e \oplus R\cdot(1-e)$. Obviously, $P = R\cdot e$ is a f.g. projective R-module, but it is not free. □

THEOREM 1.2: There is a decomposition for every semilocal ring R, that is

$$R = R_1 \oplus R_2 \oplus \dots \oplus R_n$$

where every R_i is an indecomposable semilocal ring. If P is a f.g. projective R-module, then it has a decomposition $P = R_1^{r_1} \oplus \dots \oplus R_n^{r_n}$. Therefore P is free iff $r_1 = r_2 = \dots = r_n$.

PROOF: Because the orthogonal idempotent elements are mapped into the orthogonal idempotent elements, and R/J is a semisimple artinian ring, then R/J, so R, has no infinite orthogonal idempotent elements. By the relation between ring decompositions and orthogonal idempotent elements it leads the decomposition of R, that is

$$R = R_1 \oplus R_2 \oplus \dots R_n. \qquad (2)$$

R is semilocal, so is $R_i = R/\bigoplus_{j\neq 1} R_j$. Hence R_i is an indecomposable ring.

Next, assume P to be a f.g. projective module. Then

$$P = P_1 \oplus P_2 \oplus \dots \oplus P_n.$$

where $P_i = P_i\cdot R = e_i\cdot R$, $i = 1, 2, \dots, n$, and $1 = e_1 + \dots + e_n$ is the orthogonal idempotent decomposition of the identity. From the dual basis lemma, it follows that P_i is a f.g. projective module (as R_i-module). By Theorem 1.1, we have that $P_i \simeq (R_i)^{r_i}$ for some integer $r_i \geq 0$, and that

$$P = R_1^{r_1} \oplus R_2^{r_2} \oplus \dots \oplus R_n^{r_n}.$$

If $r_1 = r_2 = \dots = r_n = r$, then $P = (R_1 \oplus \dots \oplus R_n)^r = R^r$ is free. The other direction is simple. □

By Theorem 1.2, we can get the following.

COROLLARY: Let R be a semilocal ring, P a f.g. projective R-module. Then the endomorphism ring $End(_RP)$ can be expressed as

$$End(_RP) = M_{r_1}(R_1) \oplus \dots \oplus M_{r_m}(R_m),$$

where R_i is an indecomposable semilocal ring, $M_{r_i}(R_i)$ is the matrix ring over R_i.

From now on, we will discuss the unique factorization of noetherian semilocal rings. R is called regular if it is noetherian and gl.dim $R < \infty$.

THEOREM 1.3: If R is an indecomposable regular semilocal ring, then R is a unique factorization domain.

PROOF: R is a domain, because R is regular. It is well known that a noetherian domain is a unique factorization domain iff each minimal prime ideal is principal. Also we know a regular local ring is a unique factorization domain [4].

Let P be a minimal prime ideal of R. For any maximal ideal m, if $P \subset m$, then $(P)_m$ is minimal in R_m. Further, it is R_m-free. If $P \not\subset m$, then $(P)_m = (R)_m$. Hence, $(P)_m$ is R_m-projective for any maximal ideal m. It shows that P is R-projective [5]. By Theorem 1.1, P is R-free. Since R is a domain, hence P is principal and R is a unique factorization domain. □

By Theorems 1.2 and 1.3, we have the following.

COROLLARY: Every regular semilocal ring is the direct sum of finitely many unique factorization domains.

2. HOMOLOGICAL DIMENSION OF MODULES OVER SEMILOCAL RINGS

In this section, we will deal with the homological dimension of modules over semilocal rings. Except for the special expression, R denotes a noetherian semilocal ring, J denotes the Jacobson radical of R, and $\Gamma = R/J$ denotes the residual class ring. Usually, it is regarded as an R-module under the module operation $r.\bar{x} = \overline{rx}$ for $\bar{x} \in \Gamma$, $r \in R$.

<u>THEOREM 2.1</u>: Let A be a f.g. R-module. Then the following statements are equivalent:

(i) $\mathrm{Tor}_1^R(A,\Gamma) = 0$.

(ii) $\mathrm{Ext}_R^1(A,\Gamma) = 0$.

(iii) A is projective.

(iv) A is flat.

<u>PROOF</u>: (i) $\rightarrow$ (ii). If B is a simple R-module, then B is isomorphic to R/m for some maximal ideal m of R. For Γ is semisimple, and $B \simeq \Gamma/(m/J)$, so B is isomorphic to a direct summand of Γ as R-module. And Then $\mathrm{Tor}_1^R(A,B)$ is isomorphic to a direct summand of $\mathrm{Tor}_1^R(A,\Gamma)$. Hence $\mathrm{Tor}_1^R(A,B) = 0$.

Say $\Omega = Q/Z$, where Q is the rational additive group and Z is the integer ring. It is well-known that Ω is Z-injective. Moreover we know the following relation:

$$\mathrm{Tor}_1^R(A,\mathrm{Hom}_Z(\Gamma,\Omega)) \simeq \mathrm{Hom}_Z(\mathrm{Ext}_R^1(A,\Gamma),\Omega).$$

Because Γ is semisimple, it means that

$$\mathrm{Hom}_Z(\Gamma,\Omega) = \bigoplus_{\alpha \in I} B_\alpha,$$

where B_α is a simple R-module for each α. Hence

$$\bigoplus_{\alpha \in I} \mathrm{Tor}_1^R(A,B_\alpha) \simeq \mathrm{Tor}_1^R(A, \bigoplus_\alpha B_\alpha) \simeq \mathrm{Hom}_Z(\mathrm{Ext}_R^1(A,\Gamma),\Omega).$$

From this, we have proved that $\mathrm{Hom}_Z(\mathrm{Ext}_R^1(A,\Gamma),\Omega) = 0$ for $\mathrm{Tor}_1^R(A,B_\alpha) = 0$ for each α. It follows by [6] that $\mathrm{Ext}_R^1(A,\Gamma) = 0$.

(ii) $\rightarrow$ (iii) Because A is f.g. it is easy for us to get an exact sequence:

$$0 \rightarrow A' \overset{\alpha}{\rightarrow} F \rightarrow A \rightarrow 0 \tag{1}$$

where F is free with finite rank, A' is a f.g. R-module. By the semisimplicity of A'/JA', we are given that $A'/JA' \simeq \oplus_{i=1}^{n} B_i$, B_i is a simple R-module. Because $\mathrm{Ext}_R^1(A,B_i) = 0$ has been proved previously, we have that

$$\mathrm{Ext}_R^1(A,A'/JA') \simeq \bigoplus_{i=1}^{n} \mathrm{Ext}_R^1(A,B_i) = 0.$$

Hence there is an exact sequence

$$0 \to \mathrm{Hom}_R(A,A'JA') \to \mathrm{Hom}_R(F,A'/JA') \to \mathrm{Hom}_R(A',A'/JA') \to 0. \tag{2}$$

Considering the natural homomorphism $\sigma:A' \to A'/JA'$, by exactness of sequence (2), there is a homomorphism $\delta:F \to A'/JA'$ such that $\sigma = \delta\alpha$.

Let's see the diagram

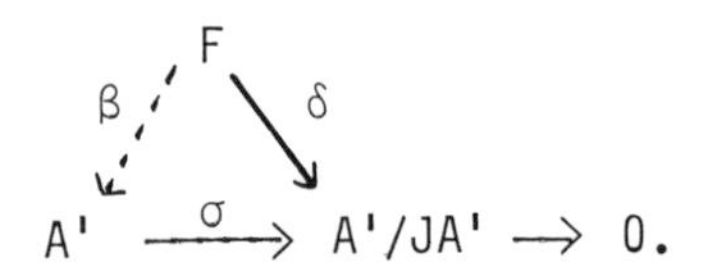

Since F is free, we have a homomorphism β such that $\sigma\beta = \delta$ and $\sigma = \sigma\beta\alpha$. Setting $\varepsilon = (1 - \beta\alpha)$, because $\sigma\varepsilon = 0$, then $\varepsilon(A') \subseteq \ker \sigma = JA'$. By the Nakayama lemma, it is clear that $JA' << A'$. From this, and $A' = \varepsilon(A') + \beta\alpha(A')$, we have that $\beta\alpha(A') = A'$ and $\beta\alpha$ is an epimorphism. It can be said that is an isomorphism, because A' is noetherian. Hence, there exists an R-isomorphism $h:A' \to A'$ such that $h\beta\alpha = 1_{A'}$. Subsequently, the exact sequence (1) is splitting, and A is projective.

(iii) → (iv) and (iv) → (i) are obvious. □

<u>COROLLARY</u>: If R is indecomposable and A is a f.g. R-module, then the following statements are equivalent:

(i) $\mathrm{Tor}_1^R(A,\Gamma) = 0$.

(ii) $\mathrm{Ext}_R^1(A,\Gamma) = 0$.

(iii) A is free.

(iv) A is flat.

THEOREM 2.2: If A is a f.g. R-module, n ($n \geq -1$) is an integer, then the homological dimension of A, $hd_R(A) \leq n$ iff $Tor_{n+1}^R(A,\Gamma) = 0$.

PROOF: It is necessary for us to prove that the condition is sufficient. Assume $Tor_{n+1}^R(A,\Gamma) = 0$. If $n = -1$, then $A \otimes_R \Gamma = 0$. From the exact sequence $A \to A/JA \to 0$, we have the exact sequence

$$A \otimes_R \Gamma \to A/JA \otimes_R \Gamma \to 0.$$

Hence, $A/JA \simeq A/JA \otimes \Gamma = A/JA \otimes_R \Gamma = 0$, and $A = JA$. By the Nakayama lemma, we can conlcude that $A = 0$, and $hd_R(A) = -1$.

If $n = 0$, namely $Tor_1^R(A,\Gamma) = 0$, A is projective by Theorem 2.1, and $hd_R(A) = 0$.

For any n (≥ 1), we can construct an exact sequence

$$0 \to A_n \to F_{n-1} \to F_{n-2} \to \dots \to F_0 \to A \to 0,$$

where F_i is free with finite rank ($1 \leq i \leq n-1$), A_n is f.g.. Hence, $Tor_1^R(A_n,\Gamma) \simeq Tor_{n+1}^R(A,\Gamma) = 0$, and A_n is projective, that is $hd_R(A) \leq n$. □

COROLLARY: gl.dim $R = hd_R(\Gamma)$.

There are similar results on the functor Ext.

THEOREM 2.3: Assume A to be a f.g. R-module, n (≥ -1) an integer. Then $hd_R(A) \leq n$ iff $Ext_R^{n+1}(A,\Gamma) = 0$.

THEOREM 2.4: Let $\ell(J/J^2)$ denote the length of R-module J/J^2. Then gl.dim $R = \ell(J/J^2)$ if and only if

$$R = R_1 \oplus F_1 \oplus \dots \oplus F_t,$$

where R_1 is a regular local ring, gl. dim R_1 = gl. dim R; and F_i ($1 \leq i \leq t$) is a field, or $F_i = 0$.

PROOF: Assume gl.dim $R = \ell(J/J^2)$. Obviously R is regular. By the Corollary of Theorem 1.3, we have that

$$R = R_1 \oplus R_2 \oplus \ldots \oplus R_t,$$

where R_i is a regular semilocal domain, $1 \leqq i \leqq t$. Let J_i be the Jacobson radical of R_i. Then we have the following:

$$\ell(J/J^2) = \ell(J_1/J_1^2) + \ell(J_2/J_2^2) + \ldots + \ell(J_t/J_t^2).$$

Since gl.dim $R = \sup_{1 \leqq i \leqq t}$ gl.dim R_i, we can assume that gl.dim R_1 = gl.dim R. It can be proved that gl.dim R_1 = k.dim R_1, the Krull dimension of R_1, and k.dim $R_1 \leqq \ell(J/J^2)$, so gl.dim $R_1 \leqq \ell(J_1/J_1^2)$. From this we infer $\ell(J_i/J_i^2) = 0$, $i \geqq 2$, and gl.dim $R_1 = \ell(J_1/J_1^2)$. By the Nakayama lemma, it implies that $J_i = 0$, $i \geqq 2$. Therefore, R_i is a field $(2 \leqq i \leqq t)$. Summing up the above discussion, we have shown that

$$R = R_1 \oplus F_1 \oplus \ldots \oplus F_t,$$

where F_i is a field, and R_1 is regular semilocal, and gl.dim R_1 = gl.dim R.

Now the remaining part is only to prove that R_1 is local. To do this, set $m_1, m_2, \ldots, m_s$ to be all the maximal ideals of R_1, and $\hat{R}_1$ to be the completion ring of R_1. Then $\hat{R}_1/\hat{J}_1 \simeq R_1/J_1$, and $\hat{J}_1 \supseteq J(\hat{R}_1)$ for R_1/J_1 is semisimple. From [7], we know $\hat{J}_1 \subseteq J(\hat{R}_1)$ so $\hat{J}_1$ is the radical of $\hat{R}_1$. Let T be a simple R_1-module. Then T is isomorphic to R_1/m for some maximal ideal m of R_1, and $\hat{T} \simeq \hat{R}_1/\hat{m}$. By the completion, m is still a maximal ideal of $\hat{R}_1$ [4], and so $\hat{T}$ is a simple $\hat{R}_1$-module. According to the isomorphism $(J_1/J_1^2)^{\wedge} \simeq \hat{J}_1/\hat{J}_1^2$, we have now found that $\ell(J_1/J_1^2) = \ell(\hat{J}_1/\hat{J}_1^2)$. On the other hand, we have gl.dim $\hat{R}_1$ = gl.dim R_1 [8]; it follows that

$$\ell(\hat{J}_1/\hat{J}_1^2) = \ell(J_1/J_1^2) = \text{gl.dim } R_1 = \text{gl.dim } \hat{R}_1.$$

Also, we have that [4]

$$\hat{R}_1 = \hat{R}_{1_{m_1}} \oplus \ldots \oplus \hat{R}_{1_{m_s}}.$$

By the same discussion as above, we know that all terms are fields except for one of them, say $\hat{R}_{1_{m_1}}$. Subsequently, $(m_i)_{m_i} \subseteq [(m_i)_{m_i}]^{\wedge} = 0$, and so $(m_i)_{m_i} = 0$ $(2 \leqq i \leqq s)$. For R_1 a domain, this is impossible, unless $m_2 = \ldots = m_s = 0$. This means that R_1 is a local ring.

In the opposite direction, if $R = R_1 \oplus F_1 \oplus \ldots + F_t$, where R_1 is a regular local ring with gl.dim R_1 = gl.dim R, F_i is a field $(1 \leqq i \leqq t)$, then it is obvious that

$$\ell(J/J^2) = \ell(m_1/m_1^2) = \text{gl.dim } R_1 = \text{gl.dim } R,$$

where m_1 is the maximal ideal of R_1. □

Finally, we are going to conclude this section with an example.

PROPOSITION: For any two given positive integers t, n, we can construct a noetherian semilocal ring R which has exactly t maximal ideals and gl.dim R = n.

PROOF: Set K to be the complex field, $\Lambda = K[x_1,\ldots,x_n]$ is the n indeterminate polynomial ring over K. By [5], section 9.1, Theorem 7, we have that gl.dim Λ = n. Since gl.dim $\Lambda = \sup_{m}$ gl.dim Λ_m, where m runs through over all maximal ideals of Λ, there is a maximal ideal m_1 such that gl.dim $\Lambda_{m_1} = n$. It is well-known that K is algebraically closed, and every maximal ideal can be expressed as

$$(x_1 - \alpha_1,\ldots, x_n - \alpha_n), \quad \alpha_i \in K,$$

and then

$$m_1 = (x_1 - \alpha_{11}, x_2 - \alpha_{12},\ldots, x_n - \alpha_{1n}).$$

Setting

$$m_j = (x_1 - \alpha_{11} - j + 1, x_2 - \alpha_{12},\ldots, x_n - \alpha_{1n}) \quad (1 \leqq j \leqq t),$$

obviously m_j ($i \leq j \leq t$) are t different maximal ideals.

Setting

$$S = \{f(x_1, x_2, \ldots, x_n) \in \Lambda \mid f(\alpha_{11} + j - 1, \alpha_{12}, \ldots, \alpha_{1n}) \neq 0, \forall j\},$$

then S is a multiplicatively closed set including 1, but not 0. Let S_1 be the multiplicatively closed set corresponding to maximal ideal m_1. It is clear that $S \subset S_1$. Setting

$$S_1/S = \{[\frac{s_1}{s}] \mid s_1 \in S_1, s \in S\}$$

then $\Lambda_{S_1} \simeq (\Lambda_S)_{S_1/S}$, namely, $\Lambda_{m_1} \simeq (\Lambda_S)_{S_1/S}$.

Now, we will show that $R = \Lambda_S$ satisfies the requirements.

In the first place, Λ_S is indecomposable noetherian, because Λ is a noetherian domain. Hence we have the following:

$$n = \text{gl.dim}\, \Lambda \geqq \text{gl.dim}\, \Lambda_S \geqq \text{gl.dim}\, (\Lambda_S)_{S_1/S} = \text{gl.dim}\, \Lambda_{S_1} = n.$$

In the second place, the number of maximal ideals of R is equal to the number of prime ideals P that are maximal with respect to $S \cap P = \emptyset$ [1]. But $S = \bigcap_{j=1}^{t} (\Lambda - m_j)$; if P is maximal with respect to $S \cap P = \emptyset$, then $P \subset (\Lambda - S) = \bigcup_{j=1}^{t} m_j$, and $P \subseteq m_j$ for some j. By the maximal of P and $m_j \cap S = \emptyset$, it must be that $P = m_j$. This shows that $R = \Lambda_S$ has, and only has, t maximal ideals.

Summing up the above discussion, it is clear that $R = \Lambda_S$ satisfies all the requirements. Moreover, R is still an indecomposable noetherian semilocal ring. □

3. TOTAL DIMENSION OF SEMILOCAL RINGS

Let A be a finitely generated module over a regular local ring R. It is well-known that the sum of the homological dimension and codimension of A is equal to the global dimension of R. However, this result is not valid in the semilocal case. In this section, the concepts of prototal dimension of modules and total dimension of rings are introduced. They can provide more information than homological dimension only. By discussing their properties, we have obtained some structures of semilocal rings with total dimension 1, 2 and 3.

For the sake of convenience, let R denote a regular semilocal ring, $m_1, m_2, \ldots, m_t$ all maximal ideals of R, J the Jacobson radical of R. Every module considered will be f.g. (finitely generated). If A is an R-module, then m_i-Codim A denotes the maximal length of A-sequence in m_i. Let

$$\text{Codim } A = \sup_{1 \leq i \leq t} m_i \text{ Codim } A.$$

Assume $\mu_1, \ldots, \mu_s$ is a sequence in J. By [9], we have the following

$$\text{hd}_R A_i + \text{Codim}_{R_i} A_i = \text{hd}_R A_j + \text{Codim}_{R_j} A_j \qquad (1)$$

where $A_i = A/(\mu_1, \ldots, \mu_i)A$, $A_0 = A$, $0 \leq i < j \leq S$.

DEFINITION 3.1: The common value of (1) is called the prototal dimension of A ($\text{ptd}_R A$); $\sup_A \text{ptd}_R A$ is called the total dimension of R (tdim R).

Clearly, if R is a local and regular ring, then $\text{Codim}_R A$ is the codimension of A, and tdim R = gl.dim R.

PROPOSITION 3.1: Assume R has two maximal ideals at least, and gl.dim R = gl.dim $R_{m_1} \geq$ gl.dim $R_{m_2} \geq \ldots \geq$ gl.dim R_{m_k}. Then tdim R = gl.dim R + gl.dim R_{m_2}.

PROOF: Let A be a nonzero f.g. module. We will divide the proof into two cases.

(i) In the case $\text{Codim}_{R_{m_1}} A_{R_{m_1}} >$ gl.dim R_{m_2}, if $\text{hd}_R A = \sup_m \text{hd}_{R_m} A_m = \text{hd}_{R_{m_1}} A_{m_1}$, then $\text{ptd}_R A = \text{hd}_R A + \text{Codim}_R A = \text{hd}_{R_{m_1}} A_{m_1} + \sup \text{Codim}_{R_m} A_m$. Since $\text{Codim}_{R_m} A_m \leq$ gl.dim R_m, $\text{Codim}_{R_{m_1}} A_{m_1} >$ gl.dim R_{m_2}. Hence

$$\sup_{m} \mathrm{Codim}_{R_m} A_m = \mathrm{Codim}_{R_{m_1}} A_{m_1},$$

and then

$$\mathrm{ptd}_R A = \mathrm{hd}_{R_{m_1}} A_{m_1} + \sup_{m} \mathrm{Codim}_{R_m} A_m$$

$$= \mathrm{hd}_{R_{m_1}} A_{m_1} + \mathrm{Codim}_{R_{m_1}} A_m$$

$$\leqq \mathrm{gl.dim}\, R_{m_1} \leqq \mathrm{gl.dim}\, R_{m_1} + \mathrm{gl.dim}\, R_{m_2}$$

$$= \mathrm{gl.dim}\, R + \mathrm{gl.dim}\, R_{m_2}.$$

In the same case, if $\mathrm{hd}_R A = \sup_{m} \mathrm{hd}_{R_m} A_m = \mathrm{hd}_{R_{m_i}} A_{m_i}$ $(i \neq 1)$, we have that $\mathrm{hd}_R A = \mathrm{hd}_{R_{m_i}} A_{m_i} \leqq \mathrm{gl.dim}\, R_{m_1} \leqq \mathrm{gl.dim}\, R_{m_2}$ $(i \neq 1)$, and then

$$\mathrm{ptd}_R A = \mathrm{hd}_R A + \mathrm{Codim}_R A \leqq \mathrm{gl.dim}\, R_{m_2} + \mathrm{Codim}_R A$$

$$\leqq \mathrm{gl.dim}\, R_{m_2} + \mathrm{Codim}_{R_{m_1}} A_{m_1} \leqq \mathrm{gl.dim}\, R_{m_2} + \mathrm{gl.dim}\, R_{m_1}$$

$$= \mathrm{gl.dim}\, R + \mathrm{gl.dim}\, R_{m_2}.$$

(ii) In the case $\mathrm{Codim}_{R_{m_1}} A_{m_1} \leqq \mathrm{gl.dim}\, R_{m_2}$, for $\mathrm{Codim}_R A \leqq \mathrm{gl.dim}\, R_{m_2}$ we can get that $\mathrm{ptd}_R A = \mathrm{hd}_R A + \mathrm{Codim}_R A \leqq \mathrm{gl.dim}\, R + \mathrm{gl.dim}\, R_{m_2}$. In any case, we always have that

$$\mathrm{ptd}_R A \leqq \mathrm{gl.dim}\, R + \mathrm{gl.dim}\, R_{m_2}.$$

and then

$$\mathrm{tdim}\, R = \sup_{A} \mathrm{ptd}_R A \leqq \mathrm{gl.dim}\, R + \mathrm{gl.dim}\, R_{m_2}.$$

Next, set $A = R \oplus R/m_1$; subsequently, we have that

$$hd_R A = hd_R R/m_1 = gl.dim\ R.$$

Since

$$A_{m_2} = R_{m_2} \oplus (R/m_1)_{m_2} = R_{m_2},$$

therefore

$$Codim_R A = \sup_m Codim_{R_m} A_m \geq Codim_{R_{m_2}} A_{m_2}$$

$$= Codim_{R_{m_2}} R_{m_2} = gl.dim\ R_{m_2}.$$

From these, it follows that

$$ptd_R A = hd_R A + Codim_R A \geq gl.dim\ R + gl.dim\ R_{m_2}.$$

Putting the above results together, we have that

$$tdim\ R = gl.dim\ R + gl.dim\ R_{m_2}. \quad \square$$

THEOREM 3.1: tdim R = gl.dim R if and only if

$$R = R_0 \oplus F_1 \oplus \ldots \oplus F_s$$

where F_i is a field, $1 \leq i \leq s$, R_0 is a local ring, and gl.dim R_0 = gl.dim R.

PROOF: If R is a local ring, obviously the conclusion is true. Next, we assume R has two maximal ideals at least.

Since R is a regular semi-local ring, by the corollary of Theorem 1.3, there is a decomposition

$$R = R_0 \times R_1 \times \ldots \times R_s,$$

where R_i is a regular domain ($0 \leq i \leq s$). From this, we get that gl.dim R = $\max_i$ gl.dim R_i. Without loss of generality, set gl.dim R = gl.dim R_0.

Assume tdim R = gl.dim R. In the first place, we show that R_0 is local. If this is not true, there are two maximal ideals at least. Let $m_0^{(1)}, m_0^{(2)}, \ldots, m_0^{(n)}$ be the maximal ideals of R_0, and gl.dim R_0 = gl.dim $R_{0_{m_0^{(1)}}} \geq$ gl.dim $R_{0_{m_0^{(2)}}} \geq \ldots \geq$ gl.dim $R_{0_{m_0^{(n)}}}$. By Proposition 3.1, we have that tdim R_0 = gl.dim R_0 + gl.dim $R_{0_{m_0^{(2)}}}$. If gl.dim $R_{0_{m_0^{(2)}}} > 0$, then $m' = m_0^{(2)} \times R_1 \times \ldots \times R_s$ is a maximal ideal of R, and gl.dim $R_{m'}$ = gl.dim $R_{0_{m_0^{(2)}}} \leq$ gl.dim $R_{0_{m_0^{(1)}}}$ = gl.dim R_m, where $m = m_0^{(1)} \times R_1 \times \ldots \times R_s$. By Proposition 3.1 again, we get that tdim R $\geq$ gl.dim R + gl.dim $R_{m'}$ = gl.dim R + gl.dim $R_{0_{m_0^{(2)}}} >$ gl.dim R. This is contradictory to the hypothesis, tdim R = gl.dim R, and so gl.dim $R_{0_{m_0^{(2)}}} = 0$. However, since R_0 is a domain, it follows that K.dim $R_{0_{m_0^{(2)}}} = 0$, and $(m_0^{(2)})_{m_0^{(2)}} = 0$. Thus, for $y (\neq 0) \in m_0^{(2)}$, there is a $u \notin m_0^{(2)}$ such that $u.y = 0$. This is impossible because R_0 is a domain and $u \neq 0$.

In the second place, we will prove that R_i is a field $(1 \leq i \leq s)$. It is easy to see that $s > 0$, otherwise R only has one maximal ideal. Assume R_1 not to be a field. Then R_1 is a domain and gl.dim $R_1 > 0$. Let m_1' be a maximal ideal of R_1 such that gl.dim R_1 = gl.dim $R_{1m_1'}$. Then $m' = R_0 \times m_1' \times R_2 \times \ldots \times R_s$ is a maximal ideal of R. Let m_0 be the unique maximal ideal of R_0. By Proposition 3.1, and gl.dim $R_{m'}$ = gl.dim $R_{1m_1'}$ = gl.dim $R_1 \leq$ gl.dim R = gl.dim R_0 = gl.dim R_{0m_0} = gl.dim R_m, where $m = m_0 \times R_1 \times \ldots \times R_s$ is another maximal ideal of R, we have tdim R $\geq$ gl.dim R + gl.dim $R_{m'}$ = gl.dim R + gl.dim $R_1 >$ gl.dim R. This is contradictory to the hypothesis tdim R = gl.dim R. So we have proved that R_1 is a field. We can prove that $R_2, \ldots, R_s$ are fields by the same arguments.

In the opposite direction, assume m_0' to be the maximal ideal of R_0. Then $m_0 = m_0' \times F_1 \times \ldots \times F_s$ and $m_i = R_0 \times F_1 \times \ldots \times F_{i-1} \times 0 \times F_{i+1} \times \ldots \times F_s$ $(1 \leq i \leq s)$ are all the maximal ideals of R, and gl.dim R_{m_i} = gl. dim $F_i = 0$ $(1 \leq i \leq s)$. By Proposition 3.1, we have shown that tdim R = gl.dim R + 0 = gl.dim R. □

Before discussing the structures of semilocal rings with tdim R = 1, 2 and 3, we will give a result concerning R with gl.dim R = 1.

PROPOSITION 3.2: Let R be an indecomposable semilocal ring with gl.dim R = 1. Then R is a Dedekind domain, and then it is noetherian. Moreover R is a principal ideal domain (p.i.d.).

PROOF: Assume $x \in R$, and $x \neq 0$. Because $R \cdot x$ is f.g. projective, and R is indecomposable, by Theorem 1.1 it follows that $R \cdot x$ is free, and x is not a zerodivisor of R, and then R is a domain. By the hypothesis gl.dim R = 1, R is a Dedekind domain. Moreover it is noetherian. From this, we can prove that R is a principal ideal domain. □

THEOREM 3.2: tdim R = 1 if and only if $R = R_0 \times F_1 \times \ldots \times F_s$, where F_i is a field ($1 \leqq i \leqq s$), and R_0 is a regular local principal ideal domain.

PROOF: ($\rightarrow$) If R is local, then gl.dim R = tdim R = 1, and R is not a field. By Proposition 3.2, R is a p.i.d.

If R has two maximal ideals at least, then, from tdim R = 1, it follows that gl.dim R = 1 by Proposition 3.1. Hence, tdim R = gl.dim R, so by Theorem 3.1, $R = R_0 \times F_1 \times \ldots \times F_s$, where F_i is a field ($1 \leqq i \leqq s$), R_0 is a regular semilocal domain with gl.dim R_0 = gl.dim R = 1. So R_0 is a p.i.d., but not a field.

($\leftarrow$) Since R_0 is a local p.i.d., and it is not a field, so gl.dim R_0 = 1, and gl.dim R = gl.dim R_0 = 1. By the expression $R = R_0 \times F_1 \times \ldots \times F_s$, we have that gl.dim R = tdim R = 1 by Theorem 3.1. □

THEOREM 3.3: If tdim R = 2, then R is one of the following cases:

(i) $R = R_1 \oplus R_2 \oplus \ldots \oplus R_s$, where R_i is a p.i.d. ($1 \leqq i \leqq s$).

(ii) $R = R_0 \oplus F_1 \oplus \ldots \oplus F_s$, where R_0 is a local ring, and gl.dim R_0 = 2, F_i is a field ($1 \leqq i \leqq s$).

PROOF: If R is local and gl.dim R = tdim R = 2, it is clear that the conclusion is true. Next, we assume that R has two maximal ideals at least. By Proposition 3.1, we have that gl.dim R = tdim R = 2 or gl.dim R = 1.

When gl.dim $R = 1$, by Proposition 3.2 and Theorem 1.2, then R is isomorphic to a direct sum of finitely many principal ideal domains. When gl.dim R = tdim $R = 2$, from Theorem 3.1, we know that $R = R_0 \oplus F_1 \oplus \dots \oplus F_s$, R_0 is a local ring with gl.dim $R_0 = 2$, F_i is a field, $1 \leqq i \leqq s$. □

COROLLARY: If tdim $R = 2$, then R is isomorphic to a direct sum of finitely many principal ideal domains if and only if there are two maximal ideals m_1, m_2 such that gl.dim $R_{m_i} > 0$, $i = 1.2$.

THEOREM 3.4: If R is a principal ideal ring, and R has two maximal ideals such that gl.dim $R_{m_i} > 0$, $i = 1, 2$, then tdim $R = 2$.

PROOF: It is easy to prove that gl.dim $R = 1$. By Proposition 3.1, we have that

$$\text{tdim } R \geqq \text{gl.dim } R + \text{gl.dim } R_{m_2} = 1 + 1 = 2,$$

and

$$\text{tdim } R \leqq \text{gl.dim } R + \text{gl.dim } R = 1 + 1 = 2,$$

and then

$$\text{tdim } R = 2. \quad \square$$

THEOREM 3.5: If tdim $R = 3$, then $R = R_0 \oplus F_1 \oplus \dots \oplus F_s$, where R_0 is a local ring with gl.dim $R_0 = 3$, F_i is a field $(1 \leqq i \leqq s)$; or $R = R_0 \oplus R_1 \oplus \dots \oplus R_t$, where gl.dim $R_0 = 2$, R_i is a p.i.d. $(1 \leqq i \leqq t)$.

PROOF: If gl.dim $R = 3$, then the result has been obtained by Theorem 3.1. If gl.dim $R = 2$, then $R = R_0 \oplus R_1 \oplus \dots \oplus R_t$, so we can assume that gl.dim R = gl.dim $R_0 = 2$, and then R_0 has a maximal ideal m_0' such that gl.dim R_0 = gl.dim $R_{0m_0'} = 2$. Take any maximal ideal of R_i, m_i', $1 \leqq i \leqq t$. Then we have maximal ideals $m_0 = m_0' \oplus R_1 \oplus \dots \oplus R_t$, $m_i = R_0 \oplus R_1 \oplus \dots \oplus R_{i-1} \oplus m_i' \oplus R_{i+1} \oplus \dots \oplus R_t$, and gl.dim $R_{m_i} \leqq$ gl.dim R = gl.dim $R_{0m_0'}$ = gl.dim R_{m_0}.

By Proposition 3.1, we have the following

$$3 = \text{tdim}\, R \geqq \text{gl.dim}\, R + \text{gl.dim}\, R_{m_i} = 2 + \text{gl.dim}\, R_{m_i}.$$

Hence, $\text{gl.dim}\, R_{m_i} \leqq 1$, and $\text{gl.dim}\, R_{i_{m_i'}} = \text{gl.dim}\, R_{m_i} \leqq 1$. From these, we infer that $\text{gl.dim}\, R_i \leqq 1$. By Proposition 3.2, R_i is a p.i.d. $(1 \leqq i \leqq t)$. □

4. ASSOCIATED R-SEQUENCES

In this section, R denotes any associative ring with identity element; it is not necessary to require R to be commutative. We will introduce the concept of associated R-sequence which generalizes the concept of R-sequence. Using these, we can get some results for local rings and semilocal rings. Finally, we plan to discuss rings with the AR-property.

DEFINITION 4.1: Let R be a ring, $p,q \in R$. If $p \cdot R = R \cdot q$, p is not a left zerodivisor and q is not a right zerodivisor, then we say that p, q are associated regular elements.

DEFINITION 4.2: Let $p_1, \ldots, p_n \in R$. If there are elements $q_1, \ldots, q_n$ such that p_1, q_1 are associated regular elements, and $p_i + (p_1, \ldots, p_{i-1})$, $q_i + (p_1, \ldots, p_{i-1})$ are associated regular elements in $R/(p_1, \ldots, p_{i-1})$, then $p_1, \ldots, p_n$ is called an associated R-sequence, and we say that $q_1, \ldots, q_n$ is associated with $p_1, \ldots, p_n$.

Obviously, if $\mu_1, \ldots, \mu_s$ is an R-sequence, then it is an associated R-sequence, too, the sequence that is associated with $\mu_1, \ldots, \mu_s$ is itself.

Assume I to be an ideal of R, A to be a left $\bar{R}$-module, where $\bar{R} = R/I$. If we define the operation $r \cdot a = \bar{r} \cdot a$, $r \in R$, $a \in A$, then A becomes a left R-module. First, we have the following two results.

THEOREM 4.1: Let p, q be associated regular elements, $I = p \cdot R = R \cdot q$, $\bar{R} = R/I$. If A is a left $\bar{R}$-module, and $\text{l.hd}_{\bar{R}} A = n$, then $\text{l.hd}_R A \leqq n + 1$. Further, if $n = 0$, then $\text{l.hd}_R A = 1$.

THEOREM 4.2: Let $p_1, \ldots, p_n$ be an associated R-sequence. Then

$l.hd_R R/(p_1,\ldots, p_n) \leqq n$. If $n = 2$, q_1, q_2 are associated with p_1, p_2 and $p_1 \cdot p_2 = p_2 \cdot p_1$, $q_1 \cdot q_2 = q_2 \cdot q_1$, then $l.hd_R R/(p_1,p_2) = r.hd_R R/(p_1,p_2) = 2$.

The proof of these two theorems is similar to the R-sequence case. Hence it is omitted here.

DEFINITION 4.3: Let R be left noetherian, I an ideal of R. I is said to have the left AR-property if there is a positive integer n for any left ideal E of R such that $I^n \cap E \subseteq IE$. R is called a left AR-ring if every ideal of R has the AR-property.

THEOREM 4.3: Let R be left noetherian, R/J semisimple artinian, A a f.g. R-module. Then the following are equivalent:

(i) $\mathrm{Tor}^R_{n+1}(R/J,A) = 0$,

(ii) $\mathrm{Ext}^{n+1}_R(A,R/J) = 0$,

(iii) $l.hd_R(A) \leqq n$.

Moreover, if J has left AR-property, then $l.hd_R R/J = l.gl.\dim R$.

The conclusions (i)-(iii) are similar to Theorem 2.2 and Theorem 2.3, so their proofs are not written out here in detail. At the mention of the last result, because J has the AR-property, it follows from Corollary 2.2 of [11] that $l.gl.\dim R = l.hd_R(R/J)$.

LEMMA 4.1: Let R be left noetherian, p, q associated regular elements. Set $I = p \cdot R = q \cdot R$. If I is a prime ideal, and $\bigcap_{j=1}^{\infty} I^m = 0$, then R is a prime ring.

PROOF: From $I = p \cdot R = q \cdot R$, immediately we have that $I^m = p^m \cdot R = R \cdot q^m$. For any $a,b \in R$, if $a \neq 0$, $b \neq 0$ and $aRb = 0$, then $a \in I$ or $b \in I$, because I is a prime ideal, and $RaRRbR \subseteq I$. Now $\bigcap_{m=1}^{\infty} I^m = 0$, so there exists a positive integer s such that $a \in I^s$, but $a \notin I^{s+1}$. Then $a = p^s.r$, for some $r \notin I$, which implies that $0 = aRb = p^s \cdot r \cdot R \cdot b$, and $rRb = 0$ for p^s not a left zero-divisor of R. Using the same arguments, we can find an element $c \notin I$ such that $rRc = 0$. This is impossible, so we have that $a = 0$ or $b = 0$, and thus R is prime. □

We call R semilocal if R/J is semisimple.

LEMMA 4.2: Let R be noetherian semilocal and have no idempotent element except for 0 and 1. Then every finitely generated projective R-module is free.

The proof of the above lemma can be obtained by a similar method to that used in [14].

THEOREM 4.4: Let R be noetherian semilocal without any non-trivial idempotent, l.gl.dim R = 2, and let J(R) have the AR-property. Then we have the following:

(i) If P is a prime ideal, and $P \supseteq J$, then P is maximal.

(ii) If P is a prime ideal, and $P \not\supseteq J$, then $P = pR = Rq$, where p, q are associated regular elements.

(iii) If P is a prime ideal, and $P \subset J$, then R is prime. If $P \not\supseteq J$, and $\bigcap_{m=1}^{\infty} P^m = 0$, then R is prime, too.

(iv) If every non-trivial prime ideal P satisfies $\bigcap_{m=1}^{\infty} P^m = 0$ then R is prime.

PROOF: (i) Since $P \supseteq J$, $R/P \simeq (R/J)/(P/J)$, we know that R/J is a direct sum of finitely many minimal ideals for R noetherian semilocal. Because P is prime and R/P is a prime ring, so R/P is simple and P is maximal.

(ii) If $P \not\supseteq J$, then $hd_R\, R/P = 1$ ([11], Corollary 4.8); further, P is a finitely generated projective left (right) R-module. By Lemma 4.2, we get that

$${}_RP = Rx_1 \oplus \dots \oplus Rx_n, \quad Rx_i \simeq {}_RR,$$

and

$$P_R = y_1R \oplus \ldots \oplus y_mR, \quad y_jR \simeq R_R.$$

Secondly, we want to prove that P is both left and right essential in R. Assume L to be a nonzero left ideal of R such that $P \cap L = 0$. Then $P \cdot L \subset P \cap L$, and $P \cdot L = y_1L \oplus \ldots \oplus y_mL = 0$. From this we have gained that $L = 0$. It is impossible, hence ${}_RP \trianglelefteq R$. Using dim(A) to denote the uniform dimension of the module A, we have that $\dim({}_RP) = \dim(Rx_1) + \ldots + \dim(Rx_n) = n \cdot \dim({}_RR)$, and $\dim(P) = \dim({}_RR)$. This implies that $n = 1$, ${}_RP = PR$. And then $P = pR = Rq$, where p, q are associated regular elements.

(iii) If $P \subset J$, by (ii), we get that $P = pR = Rq$, where p, q are associated regular elements. Immediately, we can say R is a prime ring by Lemma 4.1. Similarly if $P \not\supseteq J$ and $\bigcap_{m=1}^{\infty} P^m = 0$, then R is a prime ring, too.

(iv) If R is not prime, by (i) and (ii), then every prime ideal of R certainly is maximal. Now R has only finitely many maximal ideals because it is semilocal. Let $P_1, \ldots, P_t$ be the maximal ideals of R. Then $J = P_1 \cap P_2 \cap \ldots \cap P_t$ is the nil radical of R. By the Levitzki theorem and the Hopkins theorem ([12] Theorems 15.22 and 15.20), we know that R is artinian. But R has no non-trivial idempotent, so R/J is a division ring. Thus R is local, and there is an n such that $J^{n+1} = 0$, $J^n \neq 0$. On the other hand, by Theorem 4.3 and the AR-property, we have that $2 = \text{l.lg.dim } R = \text{l.hd}_R J + 1$, so $\text{l.hd}_R J = 1$. Consider the exact sequence

$$0 \to F_1 \xrightarrow{\tau} F_0 \xrightarrow{\sigma} J \to 0$$

where F_0, F_1 are projective R-modules, $F_1 = \text{Ker } \tau$ and $F_1 << F_0$. Since $JF_0 = \text{Rad } F_0$, hence $F_1 \subseteq JF_0$, and then $J^nF_1 \subseteq J^{n+1}F_0 = 0$. But F_1 is a free R-module by Lemma 4.2, so $J^n = 0$, which is contradictory to the choice of n. Thus we have proved that R is a prime ring. □

REFERENCES

[1] Jacobson, N., Basic Algebra (II), W.H. Freeman, San Francisco, 1980.

[2] Anderson, F.W. and Fuller, K.R., Rings and Categories of Modules, Springer, Berlin, 1974.

[3] Auslander, M. and Bachsbaum, D.A., Homological dimension in local rings, Trans. Amer. Math. Soc., 85 (1957), 390-405

[4] Matsumura, H., Commutative Algebra (second edition), Benjamin/ Cummings, New York, 1981.

[5] Northcott, D.G., An Introduction to Homological Algebra, Cambridge University Press, Cambridge, 1960.

[6] Northcott, D.G., A First Course of Homological Algebra, Cambridge University Press, Cambridge, 1973.

[7] Atiyah, M.F. and MacDonald, I.G., Introduction to Commutative Algebra, Addison-Wesley, Reading, MA, 1969.

[8] Serre, J.P., Algèbre Locale-Multiplicités, Springer, Berlin, 1965.

[9] Samuel S.H. Young, On sum of homological dimension and codimension of modules over a semi-local ring, Nagoya Math. J. 33 (1968), 165-172.

[10] Northcott, D.G., Ideal Theory, Cambridge University Press, Cambridge, 1953.

[11] Brown, K.A., Hajarnavis, C.R. and MacEacharn, A.B., Noetherian rings of finite global dimension, Proc. Lond. Math. Soc. 44 (1982), 349-371.

[12] Goodearl, K.R., Ring Theory, Nonsingular Rings and Modules, Marcel Dekker, New York, 1976.

[13] Xu Jinzhong and Yi Zhong, Some statements about homological dimension and codimension, J. Math. Res. Expos, 5, No. 4 (1985), 1-8.

[14] Xu Jinzhong, The modules over a semi-local ring and their homological dimension, Annu. Chin. Math. Vol. 7A No. 6 (1986), 685-691.

Cheng Fuchang and Yi Zhong
Department of Mathematics
Guangxi Teachers' University
Guilin
China

Xu Jinzhong
Department of Mathematics
Suzhou University
Suzhou
China

JOHN CLARK

On eventually idempotent rings

1. INTRODUCTION

In this paper all rings are associative with identity. A left (or right) ideal K of a ring R is called *eventually idempotent* if there is a natural number n, in general depending on K, such that $K^n = K^{n+1}$. The ring R is called *eventually idempotent* if all its left ideals are eventually idempotent. Eventually idempotent rings were studied under the name of *stable rings* by Page [12] and we now present his Proposition 1, along with its proof.

PROPOSITION 1: For a ring R the following conditions are equivalent:

(i) all two-sided ideals of R are eventually idempotent;

(ii) R is eventually idempotent;

(iii) all right ideals of R are eventually idempotent.

PROOF: Suppose that (i) holds and let L be a left ideal of R. Since LR is a two-sided ideal of R, $(LR)^n = (LR)^{n+1}$ for some n. Then $L^{n+1} = (LR)^n L = (LR)^{n+1} L = L^{n+2}$ and so L is eventually idempotent. Hence (i) implies (ii).

Clearly (ii) implies (i) and the equivalence of (i) and (iii) now follows by symmetry. □

If, for the ring R, there is a natural number n such that $L^n = L^{n+1}$ for all left ideals L of R then we call the least such n the *left (idempotency) bound* for R. The *right* and *two-sided bounds* (if they exist) are defined analogously. Page gives an example of a commutative ring with its ideals linearly ordered which is eventually idempotent with no bound. Moreover he observes that the proof above shows that if one of the bounds exists for an eventually idempotent ring then so do the other two and, in this case, the two-sided bound either equals or is one less than the one-sided bounds. However, he was unable to give an example where equality did not occur and asked if the three bounds were always equal. It is our purpose here to give, for each n, a family of eventually idempotent rings having two-sided bound n but left bound n+1.

2. FULLY IDEMPOTENT RINGS

Eventually idempotent rings with left bound 1, i.e. having all left ideals idempotent, are called *fully left idempotent* rings, while those with two-sided bound 1 are called just *fully idempotent*. Both classes of rings have been investigated by several authors, e.g. Armendariz et al. [2], Courter [4] and Fisher [6]. However, there appears to be no examples in the literature of fully idempotent rings which are not fully left (or right) idempotent. Here we present two examples. For the first we need to consider fully left idempotent group rings. According to Fisher [6], the necessity part of the following result is due to N. Vanaja while the sufficiency part is due to R.L. Snider.

PROPOSITION 2: Let R be a ring and G be a group. Then the group ring R[G] is fully left idempotent if and only if (i) R is fully left idempotent, (ii) G is locally finite, and (iii) the order of each g in G is a unit in R.

Now let G be a group. A word consisting of variables x_j and elements g_k in G is denoted by $W(x_j,g_k)$. A finite system of word equations and word inequalities

$$W_i(x_j,g_k) = 1 \ (i = 1,\dots,m)$$

$$V_\ell(x_j,g_k) \neq 1 \ (i = 1,\dots,n)$$

is called *consistent* with G if there is a group extension H of G in which the system has a simultaneous solution. G is called *algebraically closed* if every finite system consistent with G already has a simultaneous solution in G.

A group G is called *universal* if it is locally finite, every finite group is embeddable in G, and any two finite isomorphic subgroups of G are isomorphic under some inner automorphism of G.

Algebraically closed groups, like universal groups, also contain a copy of every finite group so both types are rather large! There is also a plentiful supply of both - for further details, one can consult chapter 6 of Kegel and Wehrfritz [8], chapter IV of Lyndon and Schupp [9], and Macintyre and Shelah [10]. Our reason for introducing these two classes of groups is to use the following theorem of Bonvallet et al. [3]:

THEOREM 3: Let G be either an algebraically closed group or a universal group and let K be any field. Then the augmentation ideal $\omega(K[G])$ is the unique proper ideal of K[G] and K[G] is primitive.

Now, if G is algebraically closed or universal then, since G is infinite, the left annihilator of $A = \omega(K[G])$ in K[G] is zero. Hence $A^2 \neq 0$ and so, since A is the only proper ideal of K[G], $A^2 = A$. This shows that K[G] is fully idempotent. However, if we now choose K to have nonzero characteristic then there are elements g in G such that the order of g is zero in K. Thus, by Proposition 2, for such K, K[G] is not fully left idempotent. Hence we have found examples of eventually idempotent rings with two-sided bound 1 but left bound 2.

We now give a different type of example. Let F be a field of characteristic zero, let F(y) denote the field of rational functions in an indeterminate y and let S = F(y)[x] be the ring of noncommutative polynomials in an indeterminate x over F(y), subject to the relation $xy - yx = 1$. Then Robson [13] has shown that the subring $R = F + xS$ of S is a hereditary noetherian domain having, just as in the group ring examples above, a unique nonzero proper ideal A, namely $A = xS$, and A is idempotent. Hence R is a fully idempotent ring. Now let r be any regular element of R contained in A and let L denote the principal left ideal Rr. Then $L^2 = RrRr = Ar$. Thus if $L^2 = L$ we have $Rr = Ar$, a contradiction since r is regular and A is proper. Hence $L^2 \neq L$ and so R is not fully left idempotent.

3. EXAMPLES OF HIGHER BOUND

Now let R be any fully idempotent ring which is not fully left idempotent, so that $A^2 = A$ for any ideal of R but there is a left ideal L of R with $L^3 = L^2 \neq L$. Let $T_2(R)$ denote the ring of 2×2 upper triangular matrices over R. Then the ideals of $T_2(R)$ are of the form

$$U = \begin{bmatrix} A & C \\ 0 & B \end{bmatrix}$$

where A, B and C are ideals of R with $A + B \subseteq C$, so

$$U^2 = \begin{bmatrix} A^2 & AC + CB \\ 0 & B^2 \end{bmatrix} \quad \text{and} \quad U^3 = \begin{bmatrix} A^3 & A^2C + ACB + CB^2 \\ 0 & B^3 \end{bmatrix}.$$

Thus, since $A^2C + ACB + CB^2 = AC + ACB + CB = AC + CB$, we get $U^2 = U^3$. Hence $T_2(R)$ is eventually idempotent with two-sided bound 2. However, for the left ideal

$$G = \begin{bmatrix} 0 & R \\ 0 & L \end{bmatrix}$$

we have

$$G^2 = \begin{bmatrix} 0 & L \\ 0 & L^2 \end{bmatrix} \neq \begin{bmatrix} 0 & L^2 \\ 0 & L^2 \end{bmatrix} = G^3,$$

so that $T_2(R)$ has left bound 3.

Now let $T_n(R)$ denote the ring of $n \times n$ upper triangular matrices over our ring R for any $n > 2$. Then the ideals of $T_n(R)$ are of the form

$$U = \begin{bmatrix} A_{11} & A_{12} & \cdots & A_{1n} \\ 0 & A_{22} & \cdots & A_{2n} \\ \vdots & \vdots & & \vdots \\ 0 & 0 & & A_{nn} \end{bmatrix}$$

where A_{ij} are ideals of R with $A_{ij} \subseteq A_{ik}$ whenever $j < k$ and $A_{ij} \subseteq A_{kj}$ whenever $i > k$. Thus we may write

$$U = \begin{bmatrix} A & M \\ 0 & B \end{bmatrix}$$

where A is an ideal of $T_{n-1}(R)$, B is an ideal of R and M is an n-1 column vector of ideals of R (with the above inclusions). Then, since $B^2 = B$ and by induction we have $A^n = A^{n-1}$, we get

$$U^{n+1} = \begin{bmatrix} A^{n+1} & A^nM + A^{n-1}MB + \dots + AMB^{n-1} + MB^n \\ 0 & B^{n+1} \end{bmatrix}$$

$$= \begin{bmatrix} A^{n-1} & A^{n-1}M + A^{n-1}MB + \dots + AMB + MB \\ 0 & B \end{bmatrix} = \begin{bmatrix} A^{n-1} & A^{n-1}M + MB \\ 0 & B \end{bmatrix} = U^n.$$

Hence $T_n(R)$ has two-sided bound n. However, for the above left ideal L of R, if H is the left ideal of $T_n(R)$ given by

$$H = \begin{bmatrix} 0 & R & R & \dots & R & R \\ 0 & 0 & R & \dots & R & R \\ \vdots & \vdots & \vdots & & \vdots & \vdots \\ 0 & 0 & 0 & \dots & 0 & R \\ 0 & 0 & 0 & \dots & 0 & L \end{bmatrix},$$

then

$$H^n = \begin{bmatrix} 0 & 0 & 0 & \dots & 0 & L \\ 0 & 0 & 0 & \dots & 0 & L^2 \\ \vdots & \vdots & \vdots & & \vdots & \vdots \\ 0 & 0 & 0 & \dots & 0 & L^2 \\ 0 & 0 & 0 & \dots & 0 & L^2 \end{bmatrix} \neq H^{n+1}.$$

Thus $T_n(R)$ has left bound n + 1.

From above one might think that to get an example for a general n then one can also take the ring $T_2(S)$ of 2×2 upper triangular matrices over an n-1 example S. Unfortunately it is not that easy, as can be checked by trying $T_2(S)$ for our ring $S = T_2(R)$ above. However, we can use a certain subring of $T_2(S)$ instead, as we now illustrate briefly for n = 3.

Let A be the ideal $\begin{bmatrix} R & R \\ 0 & 0 \end{bmatrix}$ of $S = T_2(R)$ and T be the matrix ring $\begin{bmatrix} S & A \\ 0 & S \end{bmatrix}$. Then, for the above left ideal G of S, $K = \begin{bmatrix} 0 & A \\ 0 & G \end{bmatrix}$ is a left ideal of T with $K^3 \neq K^4$. The ideals of T are of the form $\begin{bmatrix} D & F \\ 0 & E \end{bmatrix}$ where D, E and F are ideals of S with $DA + AE \subseteq F \subseteq A$. Since $D^3 = D^2$, $E^3 = E^2$ and $DF = D^2F$, we have $D^3F + D^2FE + DFE^2 + FE^3 = D^2F + FE^2 = D^2F + DFE + FE^2$ and from this it follows that $V^3 = V^4$ for any ideal V of T. Hence T is eventually idempotent with two-sided bound 3 and left bound 4.

4. FINAL REMARKS

Page also asks if the left and right bounds coincide for an eventually idempotent ring. We have no answer to this at present.

It is easily shown that any von Neumann regular ring is fully left (and right) idempotent. In fact for a ring R satisfying a polynomial identity, R is von Neumann regular if and only if each ideal of R is idempotent, i.e. R is fully idempotent. This result was obtained independently by Armendariz and Fisher [1], Fisher and Snider [7] and Page [11]. This leads us to ask if the one-sided bounds equal the two-sided bound (when either exists) for an eventually idempotent ring satisfying a polynomial identity.

Page also proves that if R is an eventually idempotent commutative ring with Jacobson radical J then R/J is von Neumann regular. This result is not true for eventually idempotent rings in general since our examples in section 2 all have zero Jacobson radical.

A ring is called *right subdirectly irreducible* (denoted by RSI) if the intersection of all its nonzero right ideals is nonzero while a *restricted right subdirectly irreducible* ring (denoted by r-RSI) is one having all of its proper homomorphic images RSI. The ideal structure of r-RSI rings was effectively determined by Deshpande and Deshpande [5] except in the case of rings which are primitive and contain a nonzero primitive ideal. It was conjectured there that a primitive r-RSI ring exists which is not simple. It is straightforward to see that the rings of section 2 provide a positive response to this.

ACKNOWLEDGEMENT: I gratefully acknowledge that, following my talk at the Conference, the last example in section 2 was pointed out to me by Professor B.J. Muller.

REFERENCES

[1] Armendariz, E.P. and Fisher, J.W., Regular P.I.-rings, Proc. Amer. Math. Soc. 39 (1973), 247-251.

[2] Armendariz, E.P., Fisher, J.W. and Steinberg, S.A., Central localizations of regular rings, Proc. Amer. Math. Soc. 46 (1974), 315-321.

[3] Bonvallet, K., Hartley, B., Passman, D.S. and Smith, M.K. Group rings with simple augmentation ideals, Proc. Amer. Math. Soc. 56 (1976), 79-82.

[4] Courter, R., Rings all of whose factor rings are semi-prime, Canad. Math. Bull. 12 (1969), 417-426.

[5] Deshpande, M.G. and Deshpande, V.K., Rings whose proper homomorphic images are right subdirectly irreducible, Pacific J. Math. 52 (1974), 45-51.

[6] Fisher, J.W., Von Neumann regular rings versus V-rings, Ring Theory (Proc. Conf., Univ. Oklahoma, Norman, Okla., 1973), pp. 101-119; Lecture Notes in Pure and Applied Mathematics, Vol. 7, Marcel Dekker, New York, 1974.

[7] Fisher, J.W. and Snider, R.L., On the von Neumann regularity of rings with regular prime factor rings, Pacific J. Math. 54 (1974), 135-144.

[8] Kegel, O. and Wehrfritz, B., Locally Finite Groups, North-Holland, Amsterdam, 1973.

[9] Lyndon, R.C. and Schupp, P.E. Combinatorial Group Theory, Springer, Berlin, 1977.

[10] Macintyre, A. and Shelah, S., Uncountable universal locally finite groups, J. Algebra 43 (1976), 168-175.

[11] Page, A., Une caractérisation des anneaux réguliers à identité polynomiale, Séminaire d'algèbre non commutative (année 1972/1973), Deuxième partie, Exp. Nos. 8-9, Publ. Math. Orsay, No. 44 Univ. Paris XI, Orsay, 1973.

[12] Page, S.S., Stable rings, Canad. Math. Bull. 23, No. 2 (1980), 173-178.

[13] Robson, J.C., Idealizers and hereditary Noetherian prime rings, J. Algebra 22 (1972), 45-81.

J. Clark
Department of Mathematics and Statistics
University of Otago
P.O. Box 56, Dunedin
New Zealand

RADOSLAV DIMITRIĆ

Some remarks on the coslender radical

In this note we show that although the coslender radical C and the radical R are identical on the set of countable abelian groups, they are different in general and C is equal to an ordinal iteration of R. We prove that both C and R commute with nonmeasurable direct products, but do not commute with measurable products. A few results on R are proved in a stronger form and simpler way than the existing proofs; at the end, a question of representing coslender groups as ascending unions of pure coslender subgroups of smaller rank is discussed briefly.

Recall the definition of a coslender group introduced in [4] (see also [5]): an abelian group A is coslender if for every homomorphism $f:A \to Z^N$, almost all $\pi_n f = 0$ (π_n are the coordinate projections). Equivalently, A is coslender if Hom(A,Z) = 0 or if A does not contain Z as a direct summand.

EXAMPLE 1: Algebraically compact, cotorsion and indecomposable groups not isomorphic to Z are coslender. If A or B (say A) is coslender, then so is $A \otimes B$, because $\mathrm{Hom}(B \otimes A,Z) \cong \mathrm{Hom}(B,\mathrm{Hom}(A,Z)) \cong \mathrm{Hom}(B,0) = 0$. By Proposition 3 in [5], new coslender groups can be obtained from other coslender groups by taking homomorphic images, direct summands, extensions, direct sums, direct limits and nonmeasurable direct products. Yet another source of coslender groups can be obtained with the help of the following:

LEMMA 1: Let A be a torsion-free, nonfree group of countable rank and let B be a pure subgroup of the smallest rank which is not free. Then B is coslender.

PROOF: First of all, this smallest rank must be finite, for if every pure subgroup of A of finite rank is free, then A is an ascending union of pure finite-rank free subgroups and it has to be free by the known result of Hill ([13], Theorem 2). Thus let B have the prescribed properties. If, on the contrary, $B = Z \oplus B'$, then B' is free since $\mathrm{rk}B' < \mathrm{rk}B$ and this leads to a contradiction; thus indeed B must be coslender. □

Theorem 9 in [5] describes all finite-rank coslender groups as direct sums of indecomposable groups none of which is isomorphic to Z.

The class of coslender groups is huge. It may be suitable to study a more restricted class of groups that are slender and coslender at the same time.

EXAMPLE 2: Note that the definition of coslender group was created by reversing some of the arrows in the classical definition of slenderness. Subgroups of Q not isomorphic to either Z or Q are both slender and coslender. A finitely generated group is of the form *free* ⊕ *torsion* so it is coslender if and only if it is torsion. Thus, there are no finitely generated groups that are slender and coslender at the same time.

EXAMPLE 3: If we let G = Z and $\lambda = \aleph_0$ in Theorem 4.2 of [6], then we derive that there is an $\aleph_1$-free, slender and coslender group of cardinality $2^{\aleph_0}$. From Corollary 7 of [19] we infer that an $\aleph_1$-free group is slender and coslender if and only if it does not contain Z as a direct summand and does not contain a subgroup of the form Z^N.

Both coslender and slender groups are closed under extensions and direct sums (see [4], Theorem 2; [3], Lemma 35; and [9], Theorem 94.3, respectively), hence there are plenty of ways of constructing groups that are slender and coslender at the same time (for brevity's sake we can call them *fit* groups). Knowing the fundamental characterization of slender groups given by Nunke and the simple characterization of coslender groups ([4], Theorem 3), we see that a group is fit if and only if it is torsion-free, reduced, does not contain a copy of p-adic integers for any prime p, does not contain a copy of Z^N and does not contain Z as a direct summand. Because of the importance of fit groups we would need a more operative description of fit groups.

PROBLEM 1: Characterize fit groups in more compact terms than the ones just given above.

It has been shown (see [5]) that every group A contains a maximum coslender subgroup (called the coslender part of A and denoted by C(A)); moreover, the coslender part of a group is a pure, fully invariant subgroup of the group. Because of the fact that homomorphic images of coslender groups are again coslender, there is a subfunctor of the identity in the category Ab of Abelian groups

$$C: Ab \to Ab$$

defined on objects $A \in A$ as $CA = C(A)$ and morphisms $C(f: A \to B) = Cf: CA \to CB$, where $Cf = f \mid C(A)$. Call this the *coslender subfunctor* or the *coslender preradical*.

PROPOSITION 2: C is a radical and a socle, i.e. for every $A \in Ab$, $C(A/C(A)) = 0$ and $CCA = CA$.

PROOF: If $B = C(A/CA) \leqq A/CA$, then $B = D/CA$, for a group D, $0 \leqq CA \leqq D \leqq A$. From the exact sequence $0 \to CA \to D \to D/CA \to 0$, we get that D is coslender, i.e. $D = CA$, and so $B = C(A/CA) = 0$. $C^2 = C$, since CA is coslender. □

The class of coslender groups is the stabilizer (or radical) class $C = \{A \mid C(A) = A\}$ of the coslender radical (and socle) C. By the known result (see e.g. [18]) it has to be closed under homomorphic images and direct sums, but it is not closed under taking subgroups (even pure), as is shown in Example 2 in [5].

EXAMPLE 4: The annihilator (or semisimple) class $N = \{A \mid CA = 0\}$ of the radical C is closed under taking subgroups, direct products and direct sums, but not under homomorphic images. For instance $C(Z^N) = 0$ and $C(Z^N/Z^{(N)}) = Z^N/Z^{(N)}$, since $Z^N/Z^{(N)}$ is algebraically compact.

There is a radical closely related to C: if X is a class of groups and A an arbitrary group, then define (see [8], section 6)

$$R_X(A) = \bigcap_{f:A \to X \in X} \ker f$$

and specially

$$RA = \bigcap_{f:A \to Z} \ker f.$$

THEOREM 3: For every group A:

(a) $CA \leqq RA$;

(b) A is coslender if and only if $RA = A$;

(c) if A is countable, then $RA = CA$;

(d) if A is countable, then A is coslender if and only if $\nu A = A$, where $\nu A = \cap_{f:A \to X} \ker f$ (X is $\aleph_1$-free) is the Chase radical.

PROOF: (a) For every $f:A \to Z$, $f \mid CA = 0$; therefore, for every $f:A \to Z$, $CA \leqq \ker f$, so $CA \leqq RA$.

(b) By (a), if A is coslender, then $A = CA \leqq RA \leqq A$ so $A = RA$; and conversely, if $RA = A$, then, for every $f:A \to Z$, $fA = 0$, i.e. A is coslender.

(c) If A is countable, then, as in a theorem of Stein ([8], Theorem 19.3), $A = RA \oplus F$, where F is a free group. Clearly, for every $f:A \to Z$, $f \mid RA = 0$. If, on the contrary, RA were not coslender, then there would exist a nonzero homomorphism $\phi:RA \to Z$. Define $f:A = RA \oplus F \to Z$ by $f(x \oplus g) = \phi x$. Obviously $f \mid RA \neq 0$, which contradicts the definition of RA. Thus, indeed $CA = RA$, if A is countable.

(d) See [2], Proposition 1.2(d). □

It is very tempting to assume now that in general $CA = RA$, i.e. that C and R are the same radicals (this was a belief expressed by L. Fuchs in a private communication). It turns out that they are not.

Note first that if $RA = 0$, then

$$A = A/RA = A/(\cap_{f:A \to Z} \ker f) \hookrightarrow \prod_{f:A \to Z} (A/\ker f) \cong \prod Z$$

i.e. $A \hookrightarrow \prod Z$, and conversely, if $A \hookrightarrow Z^I$, then $RA = 0$ since $RZ^I = 0$. This simply means that $RA = 0$ if and only if the group A is torsionless. Note that $RA = CA$ need not imply that A is coslender.

We need the following proposition attributed to B.D. Jones (see [10], Proposition 2.29); the same claim without proof is given in [14] ((E), p.68).

PROPOSITION 4: The class of torsionless groups is not closed under extensions.

PROOF: If the class of torsionless groups were closed under extensions, then in every short exact sequence

$$E: 0 \longrightarrow Z \xrightarrow{f} A \longrightarrow Z^N \longrightarrow 0,$$

A has to be torsionless.

Let $f(1) = a \neq 0$. Then there is a $g \in \text{Hom}(A,Z)$ such that $ga \neq 0$. Let $ga = n$ and form the pushout corresponding to the multiplication by n in Z:

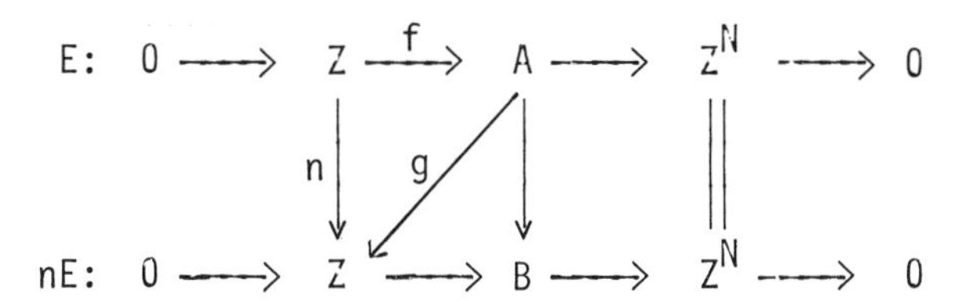

$gf(1) = n$, so $nE = 0$ ([15], Lemma 3.1, p. 72). This is a contradiction since Nunke has proved in [19] (Theorem 8) that $\text{Ext}(Z^N,Z) = \oplus_{2^c} Q \oplus \oplus_{2^c} Q/Z$, where $c = 2^{\aleph_0}$ and therefore $\text{Ext}(Z^N,Z)$ contains an element A of infinite order. □

LEMMA 5: The class $N = \{A \mid CA = 0\}$ is closed under extensions.

PROOF: By (2.8), p.70 in [18], $CA = \cap_{B \in \mathcal{B}} B$ where $\mathcal{B} = \{B \leq A \mid C(A/B) = 0\}$. If $0 \to B \to A \to D \to 0$ is an exact sequence such that $CB = C(A/B) = 0$, then $B \in \mathcal{B}$ and thus $CA \leq B$; therefore $CB = CA = 0$. □

The last two results show that there is a group A of cardinality $2^{\aleph_0}$ such that $RA \neq 0$, but since $CZ = 0$ and $CZ^N = 0$, $CA = 0$. One can show a little more, i.e. that $RA \cong Z$, and more generally, that for every group B, there is a group B' such that $RB' = B$ ([17]; see [7], Theorem 3.6). We have thus

PROPOSITION 6: There is a group A of cardinality $2^{\aleph_0}$ such that $CA = 0$ and $RA \cong Z$.

Because of $CZ^I = 0$, for every subgroup A of Z^I, also $CA = 0$, but the opposite need not be true. Simply consider again the group A from Proposition

6; as shown there $CA = 0$, but since $RA = Z$, $A \not\hookrightarrow Z^I$, by the remark above. Thus there are essential differences between R and C.

It was pointed out to me by C. Metelli that the functorial subgroup $G^*[t] = \cap_{\psi \in \Psi_t^*(G)} \ker \psi$ where $\Psi_t^*(G) = \{\psi : G \to R \mid t(R) \not\geq t\}$ for a type t and $R \leq Q$ (defined in [16]) may be related to RG for $t = z = t(Z) = (0,\ldots,0,\ldots)$. Indeed $G^*[z] = \cap_{\psi : G \to Z} \ker \psi = RG$.

On the other hand C appears to be the same as the radical R repeated a sufficient number of times:

PROPOSITION 7: For every group A there is an ordinal α such that $CA = R^\alpha A$.

PROOF: First of all, $RCA = CA$ since CA is coslender (Theorem 3(b)) and because of $CA \leq RA$ we get $CA \leq R^\beta A$ by transfinite induction on an ordinal β.

For cardinality reasons there must exist an ordinal α such that

$$A > RA > R^2 A > \ldots > R^\alpha A = R^{\alpha+1} A = \ldots .$$

Since $R(R^\alpha A) = R^\alpha A$, $R^\alpha A$ is coslender, thus $R^\alpha A \leq CA$ and by the first part we have $R^\alpha A = CA$. □

A question whether there is a radical class of torsion-free abelian groups other than P-divisible groups (P is a set of primes) closed under direct products was asked in [11] (with some answers given in [12]). It was stated also that not a single example was known of the radical class of torsion-free groups that is closed with respect to direct products.

It has been shown in Theorem 2 in [4] that the class of coslender groups is closed under nonmeasurable direct products. Depending on whether measurable cardinals exist or not, we can now contribute to answering the questions above.

THEOREM 8: For a nonmeasurable index set I, and a family of groups A_i $(i \in I)$

$$C \prod_{i \in I} A_i = \prod_{i \in I} CA_i .$$

PROOF: Note first that for every subfunctor of the identity F, $F(\Pi_{i\in I}A_i) \leq \Pi_{i\in I} FA_i$, for any index set I. Namely if $\pi_i: \Pi_{i\in I} A_i \to A_i$ are the projections, then $F\pi_i: F(\Pi_{i\in I} A_i) \to FA_i$ are their restrictions, thus $\forall a \in F(\Pi_{i\in I} A_i)$ $\forall i \in I$, $F\pi_i a \in FA_i$ and hence $a \in \Pi_{i\in I} FA_i$ (see also [1], Corollary 1.4). On the other hand, $\Pi_{i\in I} CA_i$ is coslender as a nonmeasurable product of coslender groups and therefore it has to be contained in $C\,\Pi_{i\in I} A_i$. This establishes the desired equality. □

THEOREM 9: R does not commute with measurable direct products. Specially measurable products of coslender groups need not be coslender; therefore the equality in Theorem 8 does not hold for measurable index sets.

PROOF: For a cardinal κ; let $Z_\kappa = Z^\kappa/Z^{(\kappa)}$, where Z^κ as usual denotes the κ-product of integers and $Z^{(\kappa)}$ is its subgroup - the κ-direct sum of integers. From a known result of Loś it follows that for every nonmeasurable cardinal κ, $RZ_\kappa = Z_\kappa$. This, by Theorem 3 means that Z_κ is coslender, for $\kappa < \aleph_m$ - the first measurable cardinal. However we will show that

$$R \prod_{\alpha\in O} Z_\alpha \neq \prod_{\alpha\in O} Z_\alpha = \prod_{\alpha\in O} RZ_\alpha, \qquad (*)$$

where O is the set of all regular cardinals less than the first measurable cardinal $\aleph_m$; in fact if I is a set of measurable cardinality $\aleph_m$, then we can identify $Z^I \cong \Pi_{\alpha\in O} Z^\alpha$. We use the construction from p.29 in [3] applied to the special case of Z-modules. Define $f: Z^I \to Z$ as follows: for $x = (z_i)_{i\in I}$ and $z \in Z$, set $S_z(x) = \{i \in I \mid z_i = z\}$. Then $I = \dot{\bigcup}_{z\in Z} S_z(x)$. If $\mu: P(I) \to \{0,1\}$ is a σ-additive measure on I, then μ is $|I|$-additive (see e.g. T. Jech, *Set Theory*), which implies that there is a unique $z \in Z$ such that $\mu S_z(x) = 1$; define $f(x) = z$.

Now f is a homomorphism since $\forall x,y \in Z^I$, $S_z(x) \cap S_w(y) \subset S_{z+w}(x + y)$ and if $1 = \mu S_z(x) = \mu S_w(y)$, then $\mu(S_z(x) \cap S_w(y)) = 1$. Also if $n \in Z$, then $S_z(x) \subset S_{nz}(nx)$.

Also f is an onto function since $\forall z \in Z$, $S_z(z,z,\ldots) = I$ and thus $f(z,z,\ldots) = z$. If $x \in Z^{(I)}$, then $\forall z \neq 0$, $S_z(x)$ is finite and thus $\mu S_z(x) = 0$, giving $f(x) = 0$.

The just constructed homomorphism $f: \Pi_{\alpha\in O} Z^\alpha \to Z$ induces a homomorphism

$\bar{f}:\Pi_{\alpha\in O} Z^{\alpha}/Z^{(\alpha)} \to Z$ given by $\bar{f}(X_{\alpha} + Z^{(\alpha)})_{\alpha\in O} = f(X)$, where $X = (X_{\alpha})_{\alpha\in O} \in \Pi_{\alpha\in O} Z^{\alpha}$. We need $f(\Pi_{\alpha\in O} Z^{(\alpha)}) = 0$ to show that $\bar{f}$ is well-defined. If $X \in \Pi_{\alpha\in O} Z^{(\alpha)}$, then for $z \neq 0$, $S_z(X) = \dot{\cup}$ *finite sets*. Since μ is $|I|$- additive, $\mu S_z(X) = \Sigma_I \mu$ (*finite sets*)$=0$. Thus $f(X) = 0$. So $\bar{f}$ is a nonzero homomorphism, therefore $\ker f < \Pi_{\alpha\in O} Z_{\alpha}$, proving the inequality (*). □

We also have a result of the same kind as Theorem 9, involving $R_{\mathcal{X}}$ - the radical defined earlier on:

PROPOSITION 10: If $\mathcal{X}$ is the class of groups X with the property that for every nonmeasurable set of groups $\{A_i\}_{i\in I}$, $\mathrm{Hom}(\Pi_{i\in I} A_i/\oplus_{i\in I} A_i, X) = 0$ (such groups are sometimes called cotorsion-free), then

$$R_{\mathcal{X}} \prod_{i\in I} A_i = \prod_{i\in I} R_{\mathcal{X}} A_i,$$

for every family of groups A_i, $i \in I$ (I is a nonmeasurable set).

PROOF: Just like in the proof of Theorem 8, we have $R_{\mathcal{X}} \Pi_{i\in I} A_i \leqq \Pi_{i\in I} R_{\mathcal{X}} A_i$. Assume that $f:\Pi_{i\in I} A_i \to X$ is any homomorphism. By definition

$$\prod_{i\in I} R_{\mathcal{X}} A_i = \prod_{i\in I} \bigcap_{f_i:A_i \to X\in\mathcal{X}} \ker f_i$$

and $f|\oplus_{i\in I}(\cap_{f_i:A_i\to X\in\mathcal{X}} \ker f_i) = 0$, since $f|A_i = f_i: A_i \to X$, $\forall i \in I$. This, by definition of $\mathcal{X}$ implies that $f = 0$, which in turn implies that $\Pi_{i\in I} R_{\mathcal{X}} A_i \leqq R_{\mathcal{X}} \Pi_{i\in I} A_i$. □

The last problem we would like to address here is that of representing coslender groups as smooth ascending unions of pure coslender subgroups of smaller rank. We recall a result from [5] (Proposition 10):

PROPOSITION 11: For a regular cardinal κ and a group A of rank κ, the following are equivalent:

(a) A may be represented as a smooth ascending union of pure coslender subgroups of rank $< \kappa$;

(b) For every $a \in A$, there is a pure coslender subgroup C of A with $a \in C$ and $\mathrm{rk}C < \kappa$.

COROLLARY 12: If A is separable torsion-free group of a regular infinite rank, then A is coslender if and only if it is an ascending smooth union of coslender pure subgroups of smaller rank.

PROOF: It is enough to check (b) from Proposition 11. By separability of A, every $a \in A$ is contained in a finite-rank completely decomposable direct summand of A. This summand is coslender since A is coslender. □

Considering the group from Example 3 which is $\aleph_1$-free, slender, coslender or cardinality $2^{\aleph_0}$, we see that the group cannot be represented as a smooth ascending union of pure coslender subgroups of smaller rank (compare also the comment after Proposition 10 in [5]). On the other hand, I believe in the positive answer to the following:

PROBLEM 2: Is every *countable* torsion-free group coslender if and only if it is an ascending union of pure finite-rank coslender subgroups?

REFERENCES

[1] B. Charles, Sous-groupes fonctoriels et topologiques, Studies on Abelian Groups, Dunod, Paris and Springer, Berlin, 1968, pp. 75-92.

[2] S.U. Chase, On group extensions and a problem of J.H.C. Whitehead, Topics in Abelian Groups, Scott-Foresman, Chicago, 1963, pp. 173-193.

[3] R. Dimitrić, Slenderness in abelian categories, Dissertation, Tulane University, New Orleans, 1983.

[4] R. Dimitrić, On coslender groups, Glasnik Matematički 21,(41)(1986),327-329.

[5] R. Dimitrić and B. Goldsmith, A note on coslender groups, Glasnik Mathematički, 23 No. 2 (1988).

[6] M. Dugas and R. Göbel, On radicals and products, Pacific J. Math. 118 No. 1 (1985), 79-104.

[7] T. Fay, E. Oxford and G. Walls, Preradicals in abelian groups, Houston J. Math. 8 No. 1 (1982), 39-52.

[8] L. Fuchs, Infinite Abelian Groups, vol. I, Academic Press, New York 1970.

[9] L. Fuchs, Infinite Abelian Groups, vol. II, Academic Press, New York, 1973.

[10] B.J. Gardner, Some properties of torsion classes of abelian groups, Dissertation, University of Tasmania, Hobart, 1970.

[11] B.J. Gardner, Some closure properties on radicals of abelian groups, Pacific J. Math. 42 No. 1 (1972), 45-61.

[12] B.J. Gardner, When are radical classes of abelian groups closed under direct products, Algebraic Structures and Applications, Lecture Notes in Pure and Applied Mathematics, vol. 74, Marcel Dekker, New York, 1980, pp. 87-99.

[13] P. Hill, On the freeness of abelian groups: generalization of Pontryagin's theorem, Bull. Amer. Math. Soc. 76 (1970), 1118-1120.

[14] J.P. Jans, Rings and Homology, Holt, Rinehart and Winston, New York, 1964.

[15] S. MacLane, Homology, Springer, Berlin, 1963.

[16] C. Metelli, On type-related properties of torsionfree abelian groups, Abelian Group Theory, Honolulu Proc. 1982/83, Lecture Notes in Mathematics, vol. 1006, Springer, Berlin, 1983, pp. 253-267.

[17] R. Mines, Radicals and torsion-free groups, Seminar Notes, New Mexico State University, Las Cruces, 1971.

[18] A.P. Mishina and L.A. Skorniakov, Abelian Groups and Modules, AMS Translations, vol. 107, series 2, 1976.

[19] R.J. Nunke, Slender groups, Acta Sci. Math. Szeged. 23 (1962), 67-73.

R. Dimitrić
Department of Mathematics
University of Exeter
North Park Road
Exeter EX4 4QE

M.W. EVANS

A class of semihereditary rings

A right R-module A_R is said to be *right projectively torsion-free* (A_R is P.T.F.) if for every $a \in A$, there exist subsets $\{a_1, a_2, \ldots, a_n\} \subseteq A$ and $\{x_1, x_2, \ldots, x_n\} \subseteq R$ such that $a = \Sigma_{i=1}^{n} a_i x_i$ and for all $x = R$, if $ax = 0$ then $x_i x = 0$ for all $1 \leqq i \leqq n$ [1].

An R-module A_R is said to be f-*projective* if, for each of its finitely generated submodules, the inclusion map factors through a free module. This property lies properly between flatness and projectivity [5].

It is shown in this paper that a module A is f-projective if and only if the module of row vectors ${}^{n}A$ is an $R_{(n)}$ P.T.F. module for all n. $R_{(n)}$ is the ring of all $n \times n$ matrices over R and ${}^{n}A$ the set of row vectors over A.

An extended semihereditary ring R is a ring for which the right flat R-modules are the torsion-free class of a perfect torsion theory (or equivalently the left flat R-modules satisfy this property). These rings have been discussed in [2], [3] and [4].

Rings, defined by the property that they are extended semihereditary ring for which $Q_{max}(R)$ is a left flat epimorphic extension of R, are characterized in terms of their matrix rings and also in terms of f-projective R-modules.

1. PROJECTIVELY TORSION-FREE R-MODULES

Let S_R denote the class of right P.T.F. modules. If R is a right nonsingular ring and A_R a P.T.F. module, then A_R is right nonsingular. The following characterization is given in [1]: $A_R \in S_R$ if and only if $a \in A_R$ and $aX = 0$, for a subset $X \subseteq R$, implies that $a \in A\ell_R(X)$.

THEOREM 1.1: A_R is a P.T.F. module if and only if every cyclic submodule of A_R may be factored through a finitely generated right free R-module.

PROOF: We first assume A_R is P.T.F. Let $a \in A$. Since A_R is P.T.F. there exists $\{a_i\}_{i=1}^{n}$, and $\{x_i\}_{i=1}^{n}$, such that $a = \Sigma_{i=1}^{n} a_i x_i$ and for all $x \in R$, if $ax = 0$, then $x_i x = 0$ for all x_i. Consider the diagram

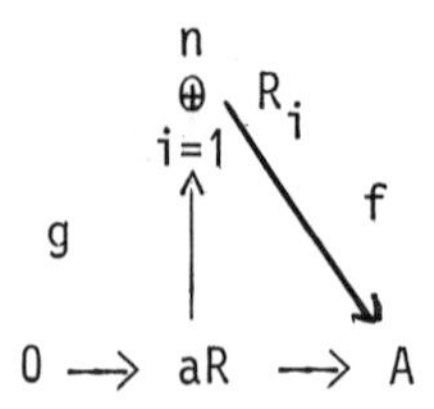

We define g by $g(a) = (x_1, x_2, \ldots, x_n)$. This is a well-defined map for if $ax = ay$ then $a(x-y) = 0$ and this implies $x_i(x-y) = 0$ for all i and hence $x_i x = x_i y$.

We define f by $f(1,0,0,\ldots,0) = a_1$, $f(0,1,\ldots,0) = a_2$, etc. Hence $f(r_1, r_2, \ldots, r_n) = \Sigma_{i=1}^{n} a_i r_i$ and in particular $gf(a) = \Sigma_{i=1}^{n} a_i x_i = a$ and $gf(ar) = ar$.

Conversely assume each cyclic submodule aR may be factored through a finitely generated free module F, i.e. there exists a commutative diagram

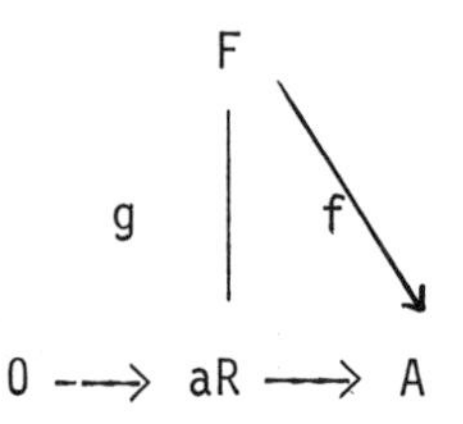

Let $g(a) = (x_1, x_2, \ldots, x_n)$. Then $f(x_1, x_2, \ldots, x_n) = \Sigma_{i=1}^{n} f(0,\ldots,1,\ldots,0)x_i$ where 1 is in the ith place and 0's in all other places. Then let $f(0,\ldots,1,\ldots,0) = a_i$ where 1 is in the ith place and 0's in all other places.

Hence $a = fg(a) = \Sigma_{i=1}^{n} a_i x_i$ and if $ax = 0$, $g(ax) = 0$ which implies $x_i x = 0$ for all i. □

From this result it is clear that every right projective R-module is a P.T.F. module.

THEOREM 1.2: A_R is an f-projective R-module if and only if ${}^nA$, the module of row vectors over A, is a P.T.F. $R_{(n)}$ modules for all $n \in \mathbb{N}$.

PROOF: The proof may proceed through Theorem 1.1 above or through the equational definition of P.T.F. □

COROLLARY 1.3: Let $\phi: R \to S$ be a ring monomorphism such that S is a right ring of quotients of R with respect to a torsion theory (T,F). Then S is a right flat epimorphism if and only if $S_{(n)}$ is a P.T.F. $R_{(n)}$ module for all n.

PROOF: If $\phi:R \to S$ is right flat epimorphism, then the induced map $\phi^{\#}:R_{(n)} \to S_{(n)}$ is a ring epimorphism and $S_{(n)}$ is a right P.T.F. $R_{(n)}$ module. This follows as, if J is a left ideal of $S_{(n)}$, then $S_{(n)}(J \cap R_{(n)}) = J$. Hence for a subset, $X \subseteq R_{(n)}$

$$\ell_{S_{(n)}}(X) = S_{(n)}(\ell_{S_{(n)}}(X) \cap R_{(n)}) = S_{(n)}\ell_{R_{(n)}}(X).$$

Conversely Theorem 1.2 gives that S is a right f-projective R-module and hence $\phi:R \to S$ is a right flat epimorphism (Proposition 2.1, [5]). □

2. SEMIHEREDITARY RINGS

DEFINITION 2.1: A ring R is said to be *right strongly Baer* if every right complement ideal of R is generated by an idempotent.

THEOREM 2.2 (Theorem 4.3, [1]): R is a right strongly Baer ring if and only if R is a right P.P. ring and Q_R, the maximal right quotient ring of R, is right P.T.F. as an R-module.

Every right continuous regular ring is a strongly Baer ring [1]. There are right strongly Baer rings which are not left strongly Baer. For example, right but not left continuous regular rings satisfy this property. Every right (left) strongly Baer ring is a Baer ring in the sense of Kaplansky.

In [3] a ring was said to be an extended semihereditary ring if $M(R)$, the left flat epimorphic hull of R, is regular and every finitely generated right R-submodule of $M(R)$ is projective. These rings may be characterized as those right semihereditary rings with Q_R right flat. Another important characterization is that they are the rings for which the right (and equivalently left) flat R-modules are the torsion-free class of a perfect torsion theory [4].

If A is an R-module and I is any set, then A^I denotes the product of card(I) copies of A viewed as an R-module.

Let $\mu_{A,I}: A \otimes R^I \to A^I$ denote the canonical map $\mu_{A,I}(a \otimes \{r_i\}) = ar_i$.

If $\mu_{A,I}$ is a monomorphism for every I, A is said to be *Mittag-Leffler*.

LEMMA 2.3: Every submodule of an f-projective R-module is f-projective if and only if R is a right semihereditary ring.

PROOF: Assume R is right semihereditary. This implies w.gl.dim $R \leq 1$ and hence every submodule of an f-projective R-module is flat.

Let A be a submodule of an f-projective B. Then as R is right semihereditary R^I is left flat and hence the sequence $0 \to A \otimes R^I \to B \otimes R^I$ is exact. From this it is easily seen that A is also right Mittag-Leffler and hence A is f-projective. The converse is clear. □

THEOREM 2.4: The following are equivalent for a ring R:

(i) R is right semihereditary and Q_R, the maximal right quotient ring of R, is the left and right flat epimorphic hull of R.

(ii) $R_{(n)}$ is a right strongly Baer ring for all n.

PROOF: (i) ⇒ (ii) Being right semihereditary is a Morita invariant property and hence $R_{(n)}$ is a right semihereditary ring. $Q_{(n)}$ is the maximal right quotient ring of $R_{(n)}$ and is a right and left P.T.F. $R_{(n)}$-module by Corollary 1.3. $R_{(n)}$ is a right P.P. ring and is right strongly Baer by Theorem 2.2.

(ii) ⇒ (i) Assume $R_{(n)}$ is right strongly Baer for all n. Then $Q_{(n)}$ is a right P.T.F. module for all n and thus Q is a right f-projective R-module. This implies that Q is the right flat epimorphic hull of R (Proposition 2.1, [5]). Thus submodules of f-projective R-modules are f-projective and Lemma 2.3 gives that R is right semihereditary. □

Rings satisfying the equivalent conditions of the above theorem are clearly extended semihereditary ring. Henceforth in this paper they will be known as *right strongly extended semihereditary rings*.

LEMMA 2.5: Let R be a right semihereditary ring. Then R is a right strongly extended semihereditary ring if and only if the injective hull of every right f-projective R-module is f-projective.

PROOF: This follows by observing $E(R_R) \cong Q$, the maximal right quotient ring of R, and Theorem 5.18 of [6]. □

THEOREM 2.6: The following are equivalent for a ring R:

(i) The f-projective R-modules are the torsion-free class of a hereditary torsion theory.

(ii) R is a right strongly extended semihereditary ring.

(iii) The f-projective R-modules and right nonsingular R-modules coincide.

PROOF: (i) ⇒ (ii) From Lemma 2.3 R is a right semihereditary ring as submodules of f-projectives will be f-projective. $E_R(R)$ is f-projective and hence Q is the right flat epimorphic hull of R (Proposition 2.2, [5]).

(ii) ⇒ (iii) Let A_R be a nonsingular R-module. Then every finitely generated nonsingular R-module is projective (Theorem 5.18, [6]). Hence every nonsingular R-module is f-projective.

(iii) ⇒ (i) The nonsingular R-modules are the torsion-free class of a hereditary torsion theory. □

EXAMPLE A: Let D be a principal ideal domain, K the classical quotient field. Let R be the ring

$$\left\{\begin{bmatrix} d_1 & q \\ 0 & d_2 \end{bmatrix} : d_1, d_2 \in R,\ q \in K\right\}.$$

Then $K_{(2)}$ the ring of 2×2 matrices over K is the maximal quotient ring of R and is the left flat epimorphic hull of R. R is right and left semi-hereditary and is right strongly extended.

EXAMPLE B: Let

$$R = \begin{bmatrix} T & T \\ 0 & T \end{bmatrix},$$

where T is a commutative self-injective regular ring. Then R is a right strongly extended semihereditary ring

$$Q = \begin{bmatrix} T & T \\ T & T \end{bmatrix}$$

is the left flat epimorphic hull of R.

EXAMPLE C: The commutative F.P.F. rings of Faith are strongly extended semihereditary rings.

EXAMPLE D: C(X), the ring of real-valued functions on an extremely disconnected completely regular space (closure of every open set is open), is a right strongly extended semihereditary ring.

The condition that R is right strongly Baer and right semihereditary is not sufficient to ensure that the ring is a right strongly extended semihereditary ring. For example, a right continuous regular ring which is not right self-injective. However, these rings do have a pleasant structure as indicated in the following.

THEOREM 2.7: For a right semihereditary ring R, the following are equivalent:

(i) M(R) is a right continuous regular ring and it is a right P.T.F. module.
(ii) R is a right strongly Baer ring.

PROOF: (i) ⇒ (ii) Let J be a right complement ideal of R. Then there exists a right complement ideal J' of M(R) such that $J' \cap R = J$ and $(J \cap R)M(R) = J'$. Hence R/J may be considered as an R-submodule of M(R)/J' which is a projective M(R) module. Now M(R)/J' is a P.T.F. R-module (Lemma 2.2, [1]), and as R is right semihereditary submodules of P.T.F. modules are P.T.F. Hence R/J is a P.T.F. module and thus projective. The sequence

$$0 \to J \to R \to R/J \to 0$$

splits and $J = fR$ for some idempotent $f^2 = f \in R$. Thus R is right strongly Baer.

(ii) ⇒ (i) Let $x \in M(R)$. Then there exists $e^2 = e \in Q$ such that $xQ = eQ$. Now $xQ \cap (1-e)Q = 0$ which implies that $xM(R) \cap ((1-e)Q \cap M(R)) = 0$. Now $((1-e)Q \cap M(R))Q = (((1-e)Q \cap R)M(R))Q = ((1-e)Q \cap R)Q = (1-e)Q$ (Lemma 4.5, [1]). Thus there exists $\{x_1, x_2, \ldots, x_n\} \subseteq (1-e)Q \cap R$ and $\{q_1, q_2, \ldots, q_n\} = Q$ such that $\Sigma_{i=1}^{n} x_i q_i = 1-e$. Thus $\Sigma_{i=1}^{n} x_i Q = (1-e)Q$.

Let J be the right ideal of M(R) with generating set $x, x_1, x_2, \ldots, x_n$. If $qJ = 0$ for some $q \in Q$, then

$$q \in \ell_Q(x) \cap \cap_{i=1}^{n} \ell_Q(x_i) = \ell_Q(e) \cap \ell_Q(1-e)$$

$$= Qe \cap Q(1-e) = 0.$$

Hence J is a finitely generated dense right ideal of M(R). By Theorem 4.2 of [2] M(R) does not contain a finitely generated dense right ideal and hence J = M(R).

Since

$$xM(R) \cap \sum_{i=1}^{n} x_i M(R) \subseteq xQ \cap \sum_{i=1}^{n} x_i Q = eQ \cap (1-e)Q$$

we have that xM(R) is a direct summand and hence xM(R) = fR for some idempotent $f \in R$. Furthermore if J is a right complement ideal of M(R), $(J \cap R)M(R) = J$ and since R is a right strongly Baer ring $J = gM(R)$ for some idempotent $g^2 = g \in R$.

Finally to see that M(R) is right P.T.F. we observe that Q is a right P.T.F. module by Theorem 2.2 and as R is a right semihereditary ring, Theorem 3.3 of [1] gives that M(R) is a right P.T.F. module. □

COROLLARY 2.8: R is a right semihereditary ring if and only if $R_{(n)}$ is a right P.P. ring for all n [9].

PROOF: A new proof of this result may be obtained through Theorem 1.2, Theorem 3.3 of [1] and Lemma 2.3. □

COROLLARY 2.9: R is a right self-injective regular ring if and only if $R_{(n)}$ is a right continuous regular ring for all n [10].

PROOF: A new proof may be obtained through Theorem 1.2, Theorem 2.4 and the observation that a regular ring is right continuous if and only if it is right strongly Baer. □

3. DECOMPOSITION

R is an extended semihereditary ring, Q the maximal right quotient ring of R. If $e^2 = e \in Q$ there exists $f^2 = f \in R$ such that $eQ = fQ$. (Theorem 2.2, [3]). This also implies that $eQe \cong fQf$ as rings.

PROPOSITION 3.1: If e is a central idempotent of R, an extended semihereditary ring, eR is an extended semihereditary ring. Furthermore if R is a right strongly extended semihereditary ring, then eR is a right strongly extended semihereditary ring.

PROOF: It is well-known that eR is both right and left semihereditary. If M(R) is the left flat epimorphic hull of R it will be shown that if e is an idempotent R, eM(R) is a left flat epimorphic extension of eR.

M(R) is a regular ring and eM(R) is a regular subring. Let $x \in eM(R)$. Then there exists a finitely generated dense right ideal I of R such that $xI \subseteq R$. Let $\{x_1, x_2, \ldots, x_n\}$ be a generating set of I in R. Then $\{ex_1, \ldots, ex_n\}$ generates a finitely generated dense ideal I' of eR and $xI' \subseteq eR$. Since eR is right semihereditary the flat epimorphic hull P, of R, has no finitely generated dense right ideal and thus $xI'P = xP$ and thus $x \in P$. Hence $eM(R) \subseteq P$ which implies $eM(R) = P$.

The second part of the proposition follows as eQ is the maximal right ring of quotients of eM(R) and eR. □

THEOREM 3.2: For a right strongly extended semihereditary ring there is a ring decomposition

$$R = S \times T$$

where S is a reduced right strongly semihereditary ring and T is a ring which contains no central idempotents e such that eT is a reduced ring.

PROOF: Let X denote the collection of those central idempotents $e \in R$ for which the ring eR is reduced. Let $f = \vee X$ in the complete Boolean algebra of idempotents of R ($B(Q) = B(R)$ and $B(Q)$ is a complete Boolean algebra). Let $\phi: R \to \prod_{e \in X} eR$. Then if $\phi(x) = 0$, $ex = 0$ for all $e \in X$ and

$$x \in r_R(x) = r_Q(x) \cap R = \bigcap_{e \in X} (1-e)Q \cap R = (\bigwedge_{e \in X} (1-e))Q \cap R$$

$$= (1 - (\vee X))R = (1-f)R.$$

Hence $(1-f)R = \ker \phi$ and fR imbeds in $\prod_{e \in X} eR$. Now $\prod_{e \in X} eR$ is a reduced ring and hence so if fR. Choose $S = fR$ which is a reduced right strongly extended semihereditary ring. Choose $T = (1-f)R$. □

PROPOSITION 3.3: If R is a reduced ring, then the following are equivalent:

(i) R is a right strongly extended semihereditary ring;

(ii) R is a right semihereditary ring and Q_R is the right and left classical quotient ring of R.

PROOF: (i) ⇒ (ii) Every finitely generated dense right ideal of R contains a nonzerodivisor (Lemma 3.5, [3]). Let $x \in Q$. Then $xI = R$ for some finitely generated dense right ideal I of R. Thus $xd = y \in R$ for d a nonzerodivisor of R. Now $dQ = Q$ and $Qd = Q$ since Q is regular and hence d is a unit of Q. Thus $x = yd^{-1}$. Similarly it can be shown that Q is a left quotient ring of R.

(ii) ⇒ (i) This is not difficult. □

Q is also the left maximal quotient ring of R since Q is strongly regular (Corollary 3.9 [6]).

Since right strongly extended semihereditary rings are Baer rings there are decompositions into types I, II and III, (see, e.g. [8]). There are of course questions arising concerning the quotient rings of these decompositions; for example, if R is a strongly extended semihereditary ring and $R = R_1 \times R_2 \times R_3$ (where R_1 is of type I, R_2 of type II and R_3 of type III), is $Q(R_i)$ a Baer ring of type I, II or III?

An interesting decomposition concerns Baer rings of type I_f. A Baer ring R is of type I_f if

(a) it contains an idempotent $0 \neq e$ such that $ef = 0$ where f is central implies $f = 0$ and eRe is a reduced Baer ring;

(b) R is directly finite, i.e. $xy = 1$ implies $yx = 1$.

PROPOSITION 3.4: Let S be a semiprime ring. S is a right strongly extended semihereditary ring of type I_f if and only if there exists right strongly extended semihereditary rings $R_1, R_2, \ldots$ such that $S \cong \Pi R_i$ and each R_i is an $i \times i$ matrix ring over some reduced right strongly extended semihereditary ring T_i.

REFERENCES

[1] M.W. Evans, Projectively torsion free modules, J. Austral. Math. Soc. (Ser. A) 20 (1975), 207-221

[2] M.W. Evans, Extensions of semi-hereditary rings, J. Austral. Math. Soc. (Ser. A) 23 (1977), 333-339.

[3] M.W. Evans, Extended semi-hereditary rings, J. Austral. Math. Soc. (Ser. A) 26 (1978), 465-474.

[4] M.W. Evans, Torsion theories over semi-hereditary rings, J. Austral. Math. Soc. (Ser. A) 40 (1986), 54-70.

[5] M. Finkel Jones, f-Projectivity and flat epimorphisms, Commun. Algebra 9, No. 16 (1981), 1603-1616.

[6] K.R. Goodearl, Ring Theory, Marcel Dekker, New York, 1976.

[7] K.R. Goodearl, Von Neumann Regular Rings, Pitman, San Francisco, 1979.

[8] I. Kaplansky, Rings of Operators, Benjamin, New York, 1968.

[9] L. Small, Semi-hereditary rings, Bull Amer. Math. Soc. 73 (1967), 656-658.

[10] Y. Utumi, On continuous rings and self-injective rings, Trans. Amer. Math. Soc. 118 (1965), 158-173.

M.W. Evans
Scotch College,
Morrison Street,
Hawthorn,
Victoria 3122
Australia

JONATHAN S. GOLAN

Embedding the frame of torsion theories in a larger context – some constructions

0. INTRODUCTION

One of the major means of studying the structure of the ring R is to make use of the set of all ideals of R. This set has the order structure of a complete modular lattice and also a related algebraic structure of a semiring. The corresponding means for studying the structure of the category R-mod of left modules over R is to make use of the frame (= complete brouwerian lattice) R-tors of all torsion theories on R-mod. This lattice does not have a corresponding semiring structure but, as we will show, can be embedded in a canonical way into the set of idempotent elements of at least two different semirings, each of which can be used to provide information on the structure of R-mod.

In what follows, all rings R are associative with multiplicative identity and all modules are unital. A (hereditary) torsion theory on the category R-mod of left modules over a ring R is an equivalence class of injective modules, two such modules being equivalent if and only if each of them can be embedded in (and hence is isomorphic to a direct summand of) a direct product of copies of the other. The study of such torsion theories harks back to Gabriel [7] and has expanded tremendously since then. The family of all such torsion theories bijectively corresponds to a set and so, without any harm, we can assume that it is a set. Moreover, operations of meet and join can be naturally defined on this family in such a manner as to make it a frame (alias complete brouwerian lattice, alias local lattice, alias locale, alias Heyting algebra). By studying the structure of this frame we can often reach conclusions about the nature of the module category over which we are working and hence about the nature of the ring R. This has been the thrust of much of my research for many years; see [8-10]. In this paper, I wish to survey some methods of embedding the frame of torsion theories in a larger ambient semiring in order to help understand it further. In particular, I will discuss two methods of embedding this frame in the set of all idempotent elements of a semiring. The first has been published in

[11]; the second is based on research by Harold Simmons in the context of arbitrary frames and on some recent joint work by Harold Simmons and me which we intend to publish in monograph form in the near future.

Notation and terminology concerning torsion theories will always follow [10]. In particular, if U is a subset of R-tors then the torsion theory $\wedge U$ is defined by the condition that a left R-module is $\wedge U$-torsion if and only if it is τ-torsion for each τ in U and the torsion theory $\vee U$ is defined by the condition that a left R-module is $\vee U$-torsionfree if and only if it is τ-torsionfree for every τ in U. If M is a left R-module then $\xi(M) = \wedge\{\tau \in \text{R-tors} \mid M \text{ is } \tau\text{-torsion}\}$ and $\chi(M) = \vee\{\tau \in \text{R-tors} \mid M \text{ is } \tau\text{-torsionfree}\}$. Then $\xi = \xi(0)$ is the unique minimal element of R-tors and $\chi = \chi(0)$ is the unique maximal element of R-tors. If τ and σ are elements of R-tors then the *pseudocomplement of* σ *relative to* τ is $(\tau:\sigma) = \vee\{\tau' \in \text{R-tors} \mid \sigma \wedge \tau' \leqq \tau\}$. This torsion theory is characterized by the condition that a left R-module M is $(\tau:\sigma)$-torsion if and only if every σ-torsion submodule of a homomorphic image of M is τ-torsion as well. The element $(\xi:\tau)$ is the unique maximal torsion theory on R-mod disjoint from τ; it is called the *pseudocomplement* of τ in R-tors and is denoted by $\tau^{\perp}$.

We say that a torsion theory τ is *below* a torsion theory σ and, following Simmons, write $\tau << \sigma$, if and only if $\tau \leqq \sigma$ and $\tau = (\tau:\sigma)$. Then $\tau << \sigma$ if and only if the following two conditions are satisfied:

(1) every τ-torsion left R-module is σ-torsion; and

(2) if M is a left R-module which is not τ-torsion then there exists a nonzero σ-torsion submodule of a homomorphic image of M which is not τ-torsion.

It is easy to see that $\tau << \chi$ for each torsion theory τ in R-mod. On the other hand, $\xi << \tau$ if and only if $\tau^{\perp} = \xi$.

1. TORSION THEORIES AND LINEAR TOPOLOGIES

A topology on R which turns R into a topological ring is uniquely determined by the family of neighborhoods of 0 in it. Such a topology is *linear* if this family of neighborhoods of 0 has a base consisting of left ideals of R. The family κ of all left ideals which are neighborhoods of 0 in a given linear topology satisfies the following three conditions:

(1) If $I \in \kappa$ then any left ideal of R containing I also belongs to κ.

(2) If $I, H \in \kappa$ then $I \cap H \in \kappa$.

(3) If $I \in \kappa$ and if $r \in R$ then $(I:r) \in \kappa$.

Conversely, any family κ of left ideals of R satisfying these three conditions is a base for the family of neighborhoods of 0 in a linear topology of R. Such a family is called a (*topologizing*) *filter* of left ideals of R. We will denote the set of all topologizing filters of left ideals of R by R-fil. If τ is a torsion theory on R-mod then the set L_τ of all left ideals I or R such that R/I is τ-torsion is a topologizing filter of left ideals which, in fact, uniquely determines τ. Thus the map $\tau \mapsto L_\tau$ defines an embedding of R-tors into R-fil.

The set R-fil has considerable structure naturally defined on it. It can surely be partially ordered by inclusion. Moreover, if Y is a subset of R-fil then $\cap Y$ also satisfies conditions (1)-(3) and so is a member of R-fil and, indeed, is the largest element of R-fil contained in each element of Y. Thus R-fil has the structure of a complete lattice of meet in which is defined by intersection. The unique minimal element of R-fil is $\eta[R] = \{R\}$ and the unique maximal element of R-fil is the set $\eta[0]$ of all left ideals of R. This lattice is not distributive, in general, as has been pointed out by Katayama [15]. Indeed, Katayama shows that R-fil is a distributive lattice if and only if the lattice of all left ideals of R is distributive, which is a very strong condition indeed. However, this lattice does have other nice properties. Thus, for example, it is *algebraic*,[4] namely it has the property that every element is the join of compact elements, where we say that a filter κ is compact if and only if $\kappa \leqq \vee Y$ implies that $\kappa \leqq \vee Y'$ for some finite subset Y' of Y. It is also true that every proper element of R-fil is contained in a coatom of R-fil. For details and proofs concerning the structure of R-fil, refer to Golan [11].

More importantly, there is another operation on R-fil with which we can work, namely that of multiplication. Following Gabriel [7], we define the product $\kappa\kappa'$ of filters κ and κ' to be the set of all left ideals I of R satisfying the condition that there exists a left ideal H of κ' such that:

(1) $I \subseteq H$; and

(2) $(I:a) \in \kappa$ for all a in H.

This operation is associative but is not, in general, commutative. The filter $\eta[R]$ acts as a multiplicative identity while the filter $\eta[0]$ acts as a zero-element: $\kappa\eta[0] = \eta[0] = \eta[0]\kappa$ for all κ in R-fil. No nontrivial element of R-fil has a multiplicative inverse.

Multiplication distributes over intersection differently on the left and on the right. If $\kappa \in$ R-fil and if Y is a nonempty subset of R-fil then $\kappa(\cap Y) = \cap\{\kappa\kappa' \mid \kappa' \in Y\}$ and $(\cap Y)\kappa \subseteq \cap\{\kappa'\kappa \mid \kappa' \in Y\}$, with equality holding when Y is finite (and in certain infinite cases as well). In particular, this shows us that (R-fil,$\cap$,$\cdot$) has the structure of a semiring with zero element. Semirings were first studied implicitly by Dedekind and explicitly by Vandiver [23]; they are described in detail in Almeida Costa [1, 2]. They have in recent years been used considerably in automata theory, optimization and theoretical computer science, and so the interest in them has revived.

As we have already noted, R-fil has no nontrivial invertible elements. However, we have the following important approximation to inversion of elements. If κ and κ' are elements of R-fil then we define the *right residual* $\kappa'^{-1}\kappa$ of κ by κ' to be the unique minimal filter κ'' in R-fil satisfying $\kappa'\kappa'' \supseteq \kappa$. Such a filter always exists. Similarly, we define the *left residual* $\kappa\kappa'^{-1}$ of κ by κ' to be the unique minimal filter κ'' in R-fil satisfying $\kappa''\kappa' \supseteq \kappa$. Such a filter does not necessarily exist but does exist if κ is of the form $\eta[I]$ for some (two-sided) ideal I of R. The general theory of residuation in ordered algebraic structures is discussed in Fuchs [6] and Blyth and Janowitz [5] and can be used to obtain many results about the structure of R-fil.

How does the image of R-tors fit into this structure? If κ is an arbitrary element of R-fil then $\kappa^2 \supseteq \kappa$. Such a filter is *idempotent* if and only if $\kappa^2 = \kappa$. As Gabriel already noted, the idempotent elements of R-fil are precisely those which come from R-tors under the above-noted embedding. Thus we see that the frame R-tors can be considered as the set of idempotent elements of the semiring R-fil. This set is not, in general, closed under multiplication in R-fil and, indeed, the product of two idempotent filters is again idempotent if and only if they commute. If $\kappa \in$ R-fil then there is a unique minimal idempotent filter κ^* containing κ. This is the unique minimal solution of the equation $X = \kappa X$ and of the equation $X = X\kappa$ in R-fil. The map $\kappa \mapsto \kappa^*$ is a closure operator on R-fil satisfying $(\kappa \cap \kappa') = \kappa^* \cap \kappa'^*$ for all κ and κ' in R-fil. Such a closure operator on a lattice is called a *nucleus* or a *modal operator*. In Fuchs [6] such operators are called *linear closure operators*.

It is sometimes possible to characterize the ring R by the way the image of R-tors sits in R-fil. Thus, for example, the strongly semiprime rings studied in Handelman [12] (and previously also considered, not under this name, by Beachy) are just those rings R for which every coatom of R-fil belongs to the image of R-tors. Semiprime left Goldie rings are examples of rings of this type. For such rings we also note that $\kappa^{-1}\eta[0] = \eta[0]\kappa^{-1}$ for every filter κ in R-fil. This question, however, is still wide open and awaits further research.

2. NUCLEI ON R-tors

We now turn to a different way of embedding the frame R-tors into a larger structure, which was developed by Simmons [16-22] in a series of papers, building on work of Beazer and Macnab [3] and Isbell [13, 14]. Their work was done for arbitrary frames, but here we will concentrate on the special case of the frame R-tors.

As defined above, a nucleus on R-tors is a closure operator f:R-tors $\to$ R-tors satisfying the additional condition that $f(\tau \wedge \sigma) = f(\tau) \wedge f(\sigma)$. Let us denote the set of all nuclei on R-tors by **N**(R-tors). We note several examples of such operators.

EXAMPLE A: If $\tau \in$ R-tors let $\mathbf{P}(\tau)$ be the set of all prime torsion theories greater than or equal to τ. Then the map $\tau \mapsto \wedge\mathbf{P}(\tau)$ is a nucleus on R-tors. More generally, if M is a left R-module then the set $\text{pinv}_\tau(M)$ of τ-*pseudo-invariants* of M consists of those elements of $\mathbf{P}(\tau)$ relative to which M is not torsion; see Golan [10] for details. The map $p_M:\tau \mapsto \wedge\text{pinv}_\tau(M)$ is a nucleus on R-tors for each left R-module M.

EXAMPLE B: A left R-module M is *decisive* if and only if, for each torsion theory τ on R-mod, M is either τ-torsion or τ-torsionfree. Each such module M defines a nucleus on q_M on R-tors defined by $q_M(\tau) = \chi$ if M is τ-torsion and $q_M(\tau) = \chi(M)$ if M is τ-torsionfree. If M is simple then $q_M = p_M$.

EXAMPLE C: An arbitrary torsion theory τ on R-tors defines three nuclei in **N**(R-tors):

(1) $u_\tau : \sigma \mapsto \sigma \vee \tau$;

(2) $v_\tau . \sigma \mapsto (\sigma : \tau)$; and

(3) $w_\tau : \sigma \mapsto (\tau : (\tau : \sigma))$.

In particular, $w_\xi : \sigma \mapsto \sigma^{\perp\perp}$ is a nucleus on R-tors.

If U is a set of nuclei on R-tors then the function $\wedge U : \sigma \to \wedge\{f(\sigma) \mid f \in U\}$ is again a nucleus on R-tors. Thus R-tors has the structure of a complete lattice, in which $f \leq g$ if and only if $f(\tau) \leq g(\tau)$ for all τ in R-tors. Indeed, **N**(R-tors) is a frame in which, for nuclei f and g, $(g:f) = \wedge\{v_{f(\tau)} g u_\tau \mid \tau \in \text{R-tors}\}$. The map $\tau \mapsto u_\tau$ is a natural embedding of R-tors into **N**(R-tors) in the category of frames. It is an isomorphism precisely when **N**(R-tors) is boolean.

With any nucleus f on R-tors we associate its *fixed set* $C[f] = \{\tau \in \text{R-tors} \mid f(\tau) = \tau\}$. Then $f(\tau) = \wedge\{\sigma \in C \mid \tau \leq \sigma\}$ for all τ in R-tors and so each nucleus is completely determined by its fixed set. A subset C of R-tors is of the form C[f] for some $f \in$ **N**(R-tors) if and only if it is closed under taking arbitrary meets and satisfies the condition that if $\tau \in C$ and $\sigma \in$ R-tors then $(\tau : \sigma) \in C$.

3. DERIVATIVES AND FILTRATIONS ON R-tors

A *derivative* on R-tors is a function d from R-tors to itself satisfying the following conditions:

(1) $\tau \leq d(\tau)$ for all τ in R-tors;

(2) if $\sigma \leq \tau$ in R-tors then $d(\sigma) \leq d(\tau)$.

Thus, for example, any nucleus on R-tors is a derivative. Many of the most important examples of derivatives are not, however, nuclei. For example:

EXAMPLE A: We have the well-known *Gabriel derivative* on R-tors defined by $d_g(\tau) = \tau \vee [\vee(\xi(M) \mid M \text{ is } \tau\text{-cocritical}\}]$. This is the same as $\vee\{\xi(M) \mid M$ is τ-artinian$\}$. A related derivative on R-tors is given by $d_n(\tau) = \vee\{\xi(M) \mid M$ is τ-noetherian$\}$.

EXAMPLE B: The *Cantor-Bendixson-Simmons derivative* on R-tors is defined by $d_{cb} = \wedge\{\sigma \mid \tau << \sigma\}$. Note that it is not necessarily the case that $\tau << d_{cb}(\tau)$ for all torsion theories τ.

EXAMPLE C: The *Boyle derivative* on R-tors is defined by $d_b(\tau) = \tau\vee[\vee\{\xi(M) \mid M$ is τ-full$\}]$, where a left R-module M is *τ-full* if and only if it is τ-torsion-free and a submodule of M is τ-dense in M when and only when it is large there.

EXAMPLE D: The *socle derivative* on R-tors is defined by $d_s(\tau) = \tau\vee\{\sigma \in$ R-tors$\mid\sigma$ is an atom over $\tau\}$.

EXAMPLE E: The *jansian hull derivative* on R-tors is defined by $d_{jh}(\tau) = \wedge\{\sigma \geqq \tau \mid \sigma$ jansian$\}$. Similarly, the *stable hull derivative* is defined by $d_{sh}(\tau) = \wedge\{\sigma \geqq \tau \mid \sigma$ stable$\}$. Note that both of these functions are closure operators on R-tors, but do not, in general, satisfy the linearity condition.

Derivatives can be transfinitely iterated. If d is a derivative on R-tors and if i is an ordinal then we define the derivative d^i inductively as follows:

(1) $d^0(\tau) = \tau$;

(2) if $i > 0$ is not a limit ordinal then $d^i(\tau) = d(d^{i-1}(\tau))$;

(3) if $i > 0$ is a limit ordinal then $d^i(\tau) = \vee\{d^h(\tau) \mid h < i\}$.

The transfinite ascending chain $\tau \leqq d(\tau) \leqq d^2(\tau) \leqq \ldots$ is called the *filtration* of the torsion theory τ with respect to the given derivative. There must be a least ordinal i such that $d^i(\tau) = d^k(\tau)$ for all $k \geqq i$. This ordinal is is called the d-*length* of τ. Since R-tors is a set, there is a least ordinal h greater than or equal to the d-length of every torsion theory. We denote the derivative d^h by d^∞.

Let **D**(R-tors) be the set of all derivatives on R-tors. If W is a nonempty subset of **D**(R-tors) then we set $\wedge W$ be the function $\tau \mapsto \wedge\{d(\tau) \mid d \in W\}$. This is again a derivative on R-tors and so **D**(R-tors) is a complete lattice, containing **N**(R-tors) as a subset (but not a sublattice since the joins are different). In particular, **D**(R-tors) is a partially ordered set.

Simmons and I have spent considerable time looking at the structure of the lattice $\mathbf{D}$(R-tors). For example, it is always true that $d_{cb} \geq d_b \geq d_g$ and that $d_g^\infty \geq d_n \geq d_g$. If R is a left semistable ring (i.e. if every indecomposable injective left R-module is decisive), then $d_s \geq d_g$. If R is left semistable and left seminoetherian then $d_s = d_g = d_b = d_{cb}$. One of Simmons' major results is that $d_g = d_g^\infty \wedge d_{cb}$, from which we can conclude that $d_g = d_n \wedge d_{cb}$.

The main use of derivatives is for defining dimension functions on R-mod. Indeed, if $d \in \mathbb{D}$(R-tors) and if $\tau \in$ R-tors then a left R-module M is said to have (τ,d)-*dimension* i if and only if M is $d^i(\tau)$-torsion but not $d^h(\tau)$-torsion for all $h < i$. Such dimension functions are studied in detail in Golan [9], and we will go into them no further here, except to note that every "reasonable" notion of dimension in module categories seems indeed to arise in this manner.

4. PRENUCLEI ON R-tors

If d is a derivative on R-tors satisfying the condition that $d(\tau \wedge \sigma) = d(\tau) \wedge d(\sigma)$ for all torsion theories τ and σ, then d is not necessarily a nucleus, since it still may not be a closure operator. We will call such derivatives *prenuclei* on R-tors, and denote the set of all prenuclei on R-tors by $\mathbf{P}$(R-tors). This is a subset of $\mathbf{D}$(R-tors) containing $\mathbf{N}$(R-tors) which is closed under taking arbitrary meets and, more importantly, closed under composition as well. The derivatives d_b, d_g and d_n are all prenuclei. The Cantor-Bendixson derivative is not, nor are the derivatives d_{jh} and d_{sh}.

Since $\mathbf{P}$(R-tors) is closed under composition, it is easily verified that $\mathbf{P}$(R-tors) is a semiring, addition in which is $\wedge$ and multiplication in which is composition. This semiring has a zero element as well, namely the prenucleus d_χ defined by $d_\chi: \tau \mapsto \chi$ for all τ in R-tors. The nuclei on R-tors are precisely the idempotent elements of this semiring. Moreover, as we have already seen, there is a canonical embedding $\tau \mapsto u_\tau$ of R-tors into $\mathbf{P}$(R-tors). This map can be extended to a map from R-fil to $\mathbb{P}$(R-tors) in the following manner: for each $\kappa \in$ R-fil, let u_κ be the map from R-tors to itself given by $u_\kappa: \tau \mapsto (\kappa\tau)^*$. (Here we are identifying R-tors with the set of all idempotent elements of R-fil.) If $\tau \in$ R-tors then $\tau \subseteq \kappa\tau \subseteq (\kappa\tau)^*$ in R-fil by Proposition 3.6 of Golan [11] and so $\tau \leq u_\kappa(\tau)$. Moreover, if $\tau \leq \sigma$ in R-tors then $\kappa\tau \subseteq \kappa\sigma$ in R-fil by Corollary 3.14 of Golan [11] and so

$u_\kappa(\tau) \leq u_\kappa(\sigma)$. Therefore $u_\kappa \in \mathbf{D}$(R-tors). Finally, if $\sigma, \tau \in$ R-tors then $[\kappa(\sigma \cap \tau)]^* = (\kappa\sigma \cap \kappa\tau)^* = (\kappa\sigma)^* \cap (\kappa\tau)^*$ in R-fil by Propositions 3.13 and 5.20 of Golan [11] and so $u_\kappa(\sigma \wedge \tau) = u_\kappa(\sigma) \wedge u_\kappa(\tau)$, showing that $u_\kappa \in \mathbf{P}$(R-tors).

If κ and κ' are elements of R-fil and if $\tau \in$ R-tors then, by Propositions 3.13 and 5.20 of Golan [11], we have $[(\kappa \cap \kappa')\tau]^* = (\kappa\tau \cap \kappa'\tau)^* = (\kappa\tau)^* \cap (\kappa'\tau)^*$ and so $(u_{\kappa\cap\kappa'})(\tau) = u_\kappa(\tau) \wedge u_{\kappa'}(\tau)$. This shows that $u_{\kappa\cap\kappa'} = u_\kappa \wedge u_{\kappa'}$ in $\mathbf{P}$(R-tors). Moreover, by Propositions 3.9 and 5.25 of Golan [11] we have $[(\kappa\kappa')\tau]^* = [\kappa(\kappa'\tau)]^* = [\kappa(\kappa'\tau)^*]^* = [\kappa u_{\kappa'}(\tau)]^* = u_\kappa u_{\kappa'}(\tau)$ and so $u_{\kappa\kappa'} = u_\kappa u_{\kappa'}$ in $\mathbf{P}$(R-tors). Finally, it is clear that $u_\chi = d_\chi$, and so we have shown that the function $\kappa \mapsto u_\kappa$ is a homomorphism in the category of semirings with zero element. It is not monic, however, since, by Corollary 5.24 of Golan [11], we have $u_\kappa(\tau) = u_{\kappa^*}(\tau)$ for all torsion theories τ on R-mod and so $u_\kappa = u_{\kappa^*}$. Indeed, the kernel of this homomorphism is precisely $\{\kappa \in \text{R-fil} \mid \kappa^* = \chi\}$. Thus the homomorphism is monic precisely for the strongly semiprime rings already mentioned.

5. CONLCUDING REMARKS

What are some further directions of research? As we mentioned, Simmons' general result is that if A is an arbitrary frame then $\mathbf{N}(A)$, the set of all nuclei on A, is again a frame. Moreover, for an arbitrary frame A, $\mathbf{P}(A)$, the set of all prenuclei on A, is always a semiring. This means that, in principal, we have a transfinite sequence of semirings $\mathbf{P}$(R-tors), $\mathbf{PN}$(R-tors), $\mathbf{PN}^2$(R-tors), etc. These larger structures are hard to visualize but it might be worth considering them, or their direct limit for that matter. Where this may lead, I do not know as yet.

REFERENCES

[1] A. Almeida Costa, Sur la théorie générale des demi-anneaux, Pub. Math. Debrecen 10 (1963), 14-29.

[2] A. Almeida Costa, Cours d'Algèbre Générale, vol. III (Demi-anneaux/ Anneaux/ Algèbre Homologique/Représentations/Algèbres), Fundacao Calouste Gulbenkian, Lisbon, 1974.

[3] R. Beazer and D.S.Macnab, Modal extensions of Heyting algebras, Colloq. Math. 41 (1979), 1-12.

[4] G. Birkhoff, Lattice Theory, 3rd edn., Colloquium Publ. 25, American Mathematical Society, Providence, RI, 1973.

[5] T. S. Blyth and M.F. Janowitz, Residuation Theory, Pergamon Press, Oxford, 1972.

[6] L. Fuchs, Partially Ordered Algebraic Systems, Addison-Wesley, Reading, MA, 1963.

[7] P. Gabriel, Des catégories abéliennes, Bull. Soc. Math. Fr. 90 (1962), 323-448.

[8] J. S. Golan, Localization of Noncommutative Rings, Marcel Dekker, New York, 1975.

[9] J.S. Golan, Decomposition and Dimension in Module Categories, Marcel Dekker, New York, 1977.

[10] J.S. Golan, Torsion Theories, Longman Scientific and Technical, Harlow, 1986.

[11] J.S. Golan, Linear Topologies on a Ring: An Overview, Longman Scientific and Technical, Harlow, 1987.

[12] D. Handelman, Strongly semiprime rings, Pacific J. Math. 60 (1975), 115-122.

[13] J. R. Isbell, Atomless parts of spaces, Math. Scand. 31 (1972), 5-32.

[14] J.R. Isbell, Meet-continuous lattices, Symp. Math. 16 (1975), 41-54.

[15] H. Katayama, On the lattice of left linear topologies on a ring, Preprint, 1986.

[16] H. Simmons, A framework for topology, in A. Macintyre et al. (eds.), Logic Colloquium '77, North-Holland, Amsterdam, 1978.

[17] H. Simmons, Spaces with boolean assemblies, Colloq. Math. 43 (1980), 23-39.

[18] H. Simmons, An algebraic version of Cantor-Bendixson analysis, in B. Banaschewski (ed.), Categorical Aspects of Topology and Analysis, Lecture Notes in Mathematics, No. 915, Springer, Berlin, 1982.

[19] H. Simmons, Boolean reflections of frames and extended Cantor-Bendixson analysis, Preprint, 1986.

[20] H. Simmons, The Gabriel dimension and Cantor-Bendixson rank of a ring, Preprint, 1986.

[21] H. Simmons, Ranking techniques for modular lattices, Preprint, 1986.

[22] H. Simmons, Near-discreteness of modules and spaces as measured by Gabriel and Cantor, Preprint, 1986.

[23] H.S. Vandiver, Note on a simple type of algebra in which cancellation law of addition does not hold, Bull. Amer. Math. Soc. 40 (1934), 914-920.

J.S. Golan
Department of Mathematics
University of Haifa
31999 Haifa
Israel

N.J. GROENEWALD

On some special classes of near-rings

ABSTRACT: Given a nonzero cardinal α, a near-ring N is said to be SP(α) if α is the first cardinal such that for every nonzero element a of N there exists a subset F of N such that $|F| < \alpha + 1$ and the right annihilator of aF is zero. If $\mathcal{D}$ denotes the class of all d.g. near-rings, then it is proved that the class of all SP(α) near-rings is $\mathcal{D}$-special in the sense of Kaarli. The fact that the class of all strongly prime near-rings is a $\mathcal{D}$-special class is obtained as a corollary.

BASIC CONCEPTS: Near-rings considered will be right near-rings. Undefined notions can be found in Pilz [5]. By N_0 and N we will denote the class of all zerosymmetric near-rings and the class of all near-rings respectively.

In [6] the concept of an SP(α) ring, α a nonzero cardinal, is introduced and it is proved that the class of SP(α) rings is a special class. In this note we extend this concept to near-rings.

If N is a near-ring and $0 \neq a \in N$, then a nonempty subset X of N is said to be a *(right) insulator* for a in N if $r_N(\{rx: x \in X\}) = 0$, where $r_N(A)$ denotes the right annihilator of a subset A of N. In [2] a near-ring is defined to be *(right) strongly prime* if every nonzero element of N has a finite insulator. We have the following variant of this definition. In what follows, strongly prime and insulator will be right strongly prime and right insulator respectively.

DEFINITION: Given a nonzero cardinal α, a nonzero near-ring N is said to be SP(α) if α is the first cardinal for which every nonzero element of N has an insulator of cardinality less than $\alpha + 1$.

Note that a near-ring is strongly prime if and only if it is SP(α) for some $\alpha \leq \aleph_0$. SP(α) near-rings are prime for each α. Let $0 \neq A,B$ be ideals of N. If $0 \neq a \in N$, then it has an insulator X with $|X| < \alpha + 1$. Hence if $0 \neq b \in B$ then there exists $x \in X$ such that $axb \neq 0$, i.e. $AB \neq 0$ and, therefore, N is prime.

LEMMA 1: Given a nonzero cardinal α, let N be a zerosymmetric near-ring. If A and P are ideals of N such that N/P is SP(α), then $A/(P \cap A)$ is SP(β) for some nonzero cardinal $\beta \leqq \alpha$.

PROOF: Let $b \in C_A(A \cap P) = C_N(P) \cap A$. There exists a subset X of N, $|X| < \alpha + 1$, such that if $c \in C_N(P)$ then $bx'c \in C_N(P)$ for some $x' \in X$. Let $d \in C_N(P) \cap A$. Choose $x \in X$ such that $bxd \in C_N(P)$. By the same argument there exist a subset Y of N, $|Y| < \alpha + 1$, such that if $t \in C_N(P) \cap A$, then since N is zerosymmetric, $(bxd)yt \in C_N(P) \cap A$ for some $y \in Y$. Hence $\{xdy + (P \cap A): x \in Y\}$ is an insulator for $b + (P \cap A)$ in $A/(P \cap A)$. Hence $A/P \cap A$ is SP(β) for some $\beta \leqq \alpha$. □

COROLLARY 2: Let α be a nonzero cardinal, N a zerosymmetric SP(α) near-ring and $0 \neq I \triangleleft N$. Then I is an SP(β) near-ring for some nonzero cardinal $\beta \leqq \alpha$.

PROOF: Let P = 0 in above lemma. □

Let $\mathbf{C}_\alpha$ denote the class of all SP(α) near-rings.

LEMMA 3: If N is a zerosymmetric near-ring and $0 \neq I \triangleleft\cdot N$, i.e. I is an essential ideal of N, such that $I \in \mathbf{C}_\alpha$, then $N \in \mathbf{C}_\alpha$.

PROOF: Since I is a prime near-ring, we have $\ell(I) = \{n \in N: nI = 0\} = 0$. Let $0 \neq r \in N$ be arbitrary. Since $\ell(I) = 0$, there exists $t \in I$ such that $0 \neq rt \in I$. Now, since $I \in \mathbf{C}_\alpha$, there exists a subset Y of I, $|Y| < \alpha + 1$, such that $r_I(\{rty: y \in Y\}) = 0$. For every $0 \neq a \in N$ there exists $s \in I$ such that $as \neq 0$. Hence there exists a $y \in Y$ such that $rtyas \neq 0$. Now $rtya \neq 0$ and consequently $\{ty: y \in Y\}$ is an insulator for r in N. Therefore, N is SP(β) for some nonzero $\beta \leqq \alpha$. By Corollary 2, I is SP(γ) for some nonzero $\gamma \leqq \beta$. Hence $\alpha = \gamma = \beta$ and therefore $N \in \mathbf{C}_\alpha$. □

LEMMA 4: Let $0 \neq I \triangleleft N$ for $N \in \mathbf{C}_\alpha \cap N_0$. Then $I \in \mathbf{C}_\alpha$. Hence the class of zerosymmetric SP(α) near-rings is hereditary.

PROOF: Let I be a nonzero ideal of $N \in \mathbf{C}_\alpha \cap N_0$. We first show that $I \triangleleft\cdot N$. Let $0 \neq J \triangleleft N$. Since N is a prime near-ring $IJ \neq 0$ and because $N \in N_0$, we have $0 \neq IJ \subseteq I \cap J$. Hence $I \triangleleft \cdot N$. By Corollary 2, I is SP(β) for some $\beta \leqq \alpha$. But I is essential in N, hence N is SP(β) by Lemma 3 and so $\beta = \alpha$. Hence $I \in \mathbf{C}_\alpha$. □

Let M be a class of near-rings. M is *regular* if $0 \neq I \triangleleft N \in M$ implies that $0 \neq I/K \in M$ for some $K \triangleleft I$ and *hereditary* if $I \triangleleft N \in M$ implies $I \in M$.

A class of near-rings **R** is a *Kurosh-Amitsur radical class (KA-radical class)* if it satisfies:

(R1) **R** is closed under homomorphic images;

(R2) $\mathbf{R}(N) \in \mathbf{R}$ for all $N \in \mathcal{N}$, where $\mathbf{R}(N) = \Sigma(I_\alpha \triangleleft N: I_\alpha \in \mathbf{R})$;

(R3) $\mathbf{R}(N/\mathbf{R}(N)) = 0$ for all $N \in \mathcal{N}$.

Any regular class M determines a KA-radical class UM, the upper radical class, which is defined by:

$$UM = \{N \in \mathcal{N}: N \text{ has no nonzero homomorphic image in } M\}.$$

We will also need the concept of a Hoehnke radical. Let ρ be a mapping which assigns to each near-ring N and ideal ρN of N. Such mappings will be called ideal mappings. An ideal mapping ρ is a *Hoehnke radical* (H-*radical*) if it satisfies the following conditions:

(H1) $g(\rho N) \subseteq \rho(gN)$ for all homomorphisms $g: N \to N'$;

(H2) $\rho(N/\rho N) = 0$ for all $N \in \mathcal{N}$.

Let M be a class of near-rings. Then ρ, defined by $\rho N = \cap\{I \triangleleft N: N/I \in M\}$, is always an H-radical. Mlitz [4] has shown that if ρ is an H-radical, then $\mathbf{R}_\rho(N) = \rho N$ if and only if ρ satisfies the following two conditions:

(H3) ρ is *complete* (i.e. $\rho I = I \triangleleft N$ implies $I \subseteq \rho N$);

(H4) ρ is *idempotent* (i.e. $\rho(\rho N) = \rho N$ for all $N \in \mathcal{N}$).

In [3] Kaarli introduced the following generalization of KA-radical theory. He noticed that for some radicals ρ the properties (H3) and (H4) can be proved provided N belongs to some "good" classes of near-rings.

Unfortunately these classes were closed with respect to homomorphic images but not with respect to ideals. They were thus not suitable for developing KA-radical theory. The main example of such a class is the class of d.g. near-rings.

Let $\mathcal{D}$ be a class of near-rings, closed under homomorphic images. We call a Hoehnke radical ρ of near-rings a *$\mathcal{D}$-radical* if it satisfies conditions (H3) and (H4), *provided* $N \in \mathcal{D}$. Obviously the notion of a KA-radical is a special case of a $\mathcal{D}$-radical. This is achieved if $\mathcal{D} = N$. On the other hand, all KA-radicals are $\mathcal{D}$-radicals, for any class.

A class of near-rings $\mathbf{C}$ is called *$\mathcal{D}$-special* if it satisfies the following conditions:

(S1) $\mathbf{C}$ consists of prime near-rings;

(S2) $N \in \mathbf{C} \cap \mathcal{D}$ and $I \triangleleft N \Rightarrow I \in \mathbf{C}$;

(S3) $J \triangleleft I \triangleleft N \in \mathcal{D}$ and $I/J \in \mathbf{C} \Rightarrow J \triangleleft N$ and $N/(J:I)_N \in \mathbf{C}$ where

$(J:I)_N = \{n \in N: nI \subseteq J\}$.

Given a $\mathcal{D}$-special class $\mathbf{C}$, we put

$$\rho_{\mathbf{C}}(N) = \cap\{I: N/I \in \mathbf{C}\}.$$

As in [3] we call $\rho_{\mathbf{C}}$ a *special $\mathcal{D}$-radical.* From [3] we have that the prime radical β and the upper nil radical $\mathbf{N}$ are special $\mathcal{D}$-radicals where $\mathcal{D}$ is the class of all d.g. near-rings. In [2] we defined the strongly prime radical $\mathfrak{s}(N)$ as the intersection of all the ideals I of N such that N/I is a strongly prime near-ring.

We also proved that the strongly prime radical is a $\mathcal{D}$-special radical for $\mathcal{D}$ the class of all d.g. near-rings. In the rest of this paper $\mathcal{D}$ denotes the class of all d.g. near-rings.

We now show that if α is a nonzero cardinal then the class $\mathbf{C}_\alpha$ is $\mathcal{D}$-special. We first prove the following general lemma.

<u>LEMMA 5</u>: Let M be an essentially closed class of prime d.g. near-rings. If $I \triangleleft A \triangleleft N$, N a d.g. near-ring and $A/I \in M$, then $N/(I:A)_N \in M$.

PROOF: From [3], Theorem 4.8, we have $I \triangleleft N$ and consequently $(I:A)_N/I \ntriangleleft N/I$. It is easy to show that $(I:A)_N$ is maximal in the class $A = \{J: J \triangleleft N$ and $J \cap A \subset I\}$. From this it follows that the ideal $(I:A)_N/I$ of N/I is maximal with respect to the property $((I:A)_N/I) \cap A/I = (0)$. From [1], Proposition 2, it now follows that $A/I \triangleleft \cdot\, N/(I:A)_N$. Since M is essentially closed and $A/I \in M$, it follows that $N/(I:A)_N \in M$. □

THEOREM 6: Let α be a nonzero cardinal. Then $\mathbf{C}_\alpha$ is $\mathcal{D}$-special.

PROOF: This follows from Lemma 4 and 5 and the fact that $SP(\alpha)$ near-rings are prime for each α. □

A trivial consequence of Theorem 6 is that for each nonzero cardinal α, the class of all near-rings which are $SP(\beta)$ for some nonzero cardinal $\beta \leqq \alpha$ is also a $\mathcal{D}$-special class. Setting $\alpha = \aleph_0$ we obtain:

COROLLARY 7 ([2], Theorem 2.8): The class of all strongly prime near-rings is $\mathcal{D}$-special.

In general $\mathbf{C}_\alpha$ is not N-special. This follows from [2] where we gave a near-ring N for which condition (S3) did not hold.

If N is any near-ring, let $SP_\alpha(N) = \cap\{I \triangleleft N: N/I \in \mathbf{C}_\alpha\}$.

THEOREM 8: Let N be any d.g. near-ring and $I \triangleleft N$. Then $SP_\alpha(SP_\alpha(N)) = SP_\alpha(N)$ and $SP_\alpha(I) = SP_\alpha(N) \cap I$.

PROOF: This follows from Theorem 6 and [3], Theorems 3.3 and 3.4.

COROLLARY 9 ([2], Theorem 3.3): Let N be any d.g. near-ring and $I \triangleleft N$. Then $(\mathfrak{s}(N)) = \mathfrak{s}(N)$ and $\mathfrak{s}(I) = I \cap \mathfrak{s}(N)$.

REFERENCES

[1] T. Anderson, K. Kaarli and R. Wiegandt, Radicals and subdirect decomposition, Commun. Algebra 13 (1985), 479-494.

[2] N.J. Groenewald, Strongly prime near-rings, Proc. Edinb. Math. Soc. 31 (1988), 337-343.

[3] K. Kaarli, Special radicals of near-rings (in Russian), Tartu Riikl. Ul. Toimetised Vih. 610 (1982), 53-68.

[4] R. Mlitz, Radicals and semisimple classes of Ω-groups, Proc. Edinb. Math. Soc. 23 (1980), 37-41.

[5] G. Pilz, Near Rings, North-Holland Mathematics Studies, North-Holland, Amsterdam, 1977.

[6] J.G. Raftery, On some special classes of prime rings, Quaestiones Math. 10 (1987), 257-263.

N.J. Groenwald
Department of Mathematics
University of Port Elizabeth
P.O. Box 1600
6000 Port Elizabeth
South Africa

MELVIN HENRIKSEN

Rings with a unique regular element

1. INTRODUCTION

A Boolean ring R with identity element 1 has the property that if $1 \neq a \in R$, then a is a proper divisor of 0. In [2], Cohn showed that if S is a commutative ring with an identity element 1 that is its unique invertible element, then S is a subring of a ring with the same identity element which is its only regular element (or nondivisor of 0). A number of sufficient conditions for such a ring to be Boolean are also given in [2]; see also [10].

In this note we study rings with a unique "regular" element without assuming either the existence of an identity element or commutativity. It turns out that under any reasonable definition of the word "regular", such rings have an identity element 1, and we study their subrings that contain 1. Such a ring S has the property that if $1 \neq a \in S$, then a is in a proper completely prime ideal. There are rings with a unique invertible element without this latter property, so Cohn's characterization of subrings of a commutative ring with a unique regular element does not hold for arbitrary rings. We do not know if there is a ring with a unique regular ring that fails to be commutative.

A lot of what is shown below can be learned from a careful reading of [2], but we also derive some results about rings without nonzero nilpotent elements that appear to be new.

2. ELEMENTARY PROPERTIES OF RINGS WITH A UNIQUE REGULAR ELEMENT

Throughout R will denote a ring. If $a \in R$ then the *left* (resp. *right*) annihilator of a is given by $A_\ell(a) = \{x \in R: xa = 0\}$ (resp. $A_r(a) = \{x \in R: ax = 0\}$ and the *annihilator* of a is $A(a) = A_\ell(a) \cap A_r(a)$.

DEFINITION 2.1: An element e of R is called *left regular* if $A_r(a) = \{0\}$, *right regular* if $A_\ell(a) = \{0\}$, *regular* if $A_r(a) \cup A_\ell(a) = \{0\}$ and *weakly regular* if $A(a) = \{0\}$.

Clearly any regular element is both left and right regular, and any left or right regular element is weakly regular.

<u>DEFINITION 2.2</u>: A ring with exactly one weakly regular element is called a UR-*ring*.

One of the main results of this section is that it does not matter in this last definition if "weakly regular" is replaced by "left regular", "right regular", or "regular".

<u>LEMMA 2.3</u>: Suppose e is a weakly regular element of R. Then:

(a) e^2 and $(-e)$ are weakly regular, and

(b) if $e^2 = e$ and ex or xe = 0, then e + x is weakly regular.

<u>PROOF</u>: (a) Suppose $x \in A(e^2)$ for some $x \in R$. Then

$$e(exe) = (exe)e = 0.$$

Since e is weakly regular,

$$exe = 0.$$

Thus

$$e(ex) = (ex)e = 0$$

and the weak regularity of e yields ex = 0. Similarly xe = 0, whence x = 0. Thus e^2 is weakly regular.

That (-e) is weakly regular is clear and (a) holds.

(b) Suppose $e^2 - e = 0 = ex$. Then $e(x - xe) = 0 = (x - xe)e$, so since e is weakly regular,

$$x = xe \quad \text{and} \quad x^2 = x(ex) = 0. \tag{1}$$

Now suppose that for some $y \in R$

$$(e + x)y = 0 = y(e + x). \tag{2}$$

Multiplying each side of the first equation on the left by e yields

$$ey = 0 = xy. \tag{3}$$

Making use of the second equation in (2), (1) and (3), one obtains

$$ye + yx = 0 = ye + yxe = (y + yx)e = e(y + yx).$$

Since e is weakly regular, $y + yx = 0$, so

$$yx = -y. \tag{4}$$

Multiplying both sides of (4) on the right by x and using (1) shows that $yx = 0 = y$. So $(e + x)$ is regular, and the same conclusion is reached by a symmetric argument if $xe = 0$. Thus (b) holds. □

<u>THEOREM 2.4</u>: The following properties of a ring R are equivalent.

(a) R is a UR-ring with unique weakly regular element e.

(b) R has a unique left regular element e.

(c) R has a unique right regular element e.

(d) R has a unique regular element e.

(e) R is a UR-ring that has an identity element.

<u>PROOF</u>: Suppose (a) holds and $A_r(f) = 0$ for some $f \in R$. Then $A(f) = 0$, so $f = e$ is the only possible left regular element of R. If $ex = 0$, then by Lemma 2.3, $e + x$ is weakly regular, so $e + x = e$, whence $x = 0$. Hence (b) holds.

Similarly, (a) implies (c). Since any left or right regular element is weakly regular, (b) or (c) implies (a). Thus (a), (b) and (c) are equivalent. Since an element is regular if and only if it is both left and right regular, (d) is also equivalent to (a), (b) and (c).

Suppose once more that (a) holds. By Lemma 2.3, $e = e^2$ and $e(y - ey) = 0$

for any $y \in R$. Since (a) implies (b), e is a left identity for R. Similarly, e is also a right identity for R, so (e) holds. Clearly (e) implies (a) and the proof of the theorem is complete. □

The following definitions will facilitate the statements of many results in the sequel.

DEFINITION 2.5: Suppose R is a ring with identity element 1.

(a) If $ab = ba = 1$ for $a,b \in R$ implies $a = b = 1$, then R is called a UI-*ring*.

(b) If R is a subring of a UR-ring whose identity element is also 1, then R is called an SUR-*ring*.

Recall that a ring whose only nilpotent element is 0 is said to be *reduced*. Since $a^n = 0$ implies $(1 - a)^{-1} = (1 + a + \ldots - \ldots + a^{n-1})$, every UI-ring is reduced, and since -1 is invertible, each element of a UI-ring has characteristic 2. The following is an immediate consequence of Theorem 2.4.

COROLLARY 2.6: Every SUR-ring is a UI-ring; in particular it is reduced and $2x = 0$ for every $x \in R$.

Since $e(1 - e) = 0$ for every idempotent e in a ring with identity element, a Boolean ring with identity element is a UR-ring. As noted in the introduction, Cohn showed in [2] that every UI-ring is an SUR-ring. In particular the ring $Z_2[x]$ of polynomials over the ring of integers mod 2 is an SUR-ring that fails to be Boolean. SUR-rings will be studied in the next section. The remainder of this section is devoted to obtaining sufficient conditions for a UR-ring to be Boolean.

Recall that since whenever $e^2 = e$ and x in R, then $(ex - exe)^2 = 0 = (xe - exe)^2$, so every idempotent element of a reduced ring is in its centre.

I have been unable to find the following simple lemma in the mathematical literature.

LEMMA 2.7: If R is reduced and ab is idempotent for some $a,b \in R$, then $ab = ba$.

PROOF: Note first that $(ba)^3 = b(ab)^2a = (ba)^2$, and hence $(ba)^4 = (ba)^3 = (ba)^2$. Thus

$$[(ba)^2 - (ba)]^2 = (ba)^4 - 2(ba)^3 + (ba)^2 = 0.$$

Since R is reduced, ba is idempotent, and hence is in the centre of R. So

$$\begin{aligned}(ab - ba)^2 &= (ab)^2 - (ab)ba - (ba)ab + (ba)^2 \\ &= (ab)^2 - b(ab)a - a(ba)b + (ba)^2 \\ &= 0.\end{aligned}$$

So ab = ba since R is reduced. □

It follows that if R is reduced then A(a) is a (two-sided) ideal of R for any $a \in R$.

DEFINITION 2.8: An element a of R is said to be *regular in the sense of Von Neumann, or* VN-*regular*, if there is an $x \in R$ such that $axa = a$. A ring each of whose elements is VN-regular is called a VN-*regular ring*.

If $a \in R$ is VN-regular, there is an $x \in R$ such that

$$axa = a \quad \text{and} \quad xax = x. \tag{5}$$

To see this, note that if $ax'a = a$ and $x = x'ax'$, then (5) holds.

Observe also that ax and xa are idempotents.

A characterization of Boolean rings with identity element follows. Part (c) of this result is credited to M. Drazin in [2].

THEOREM 2.9: Suppose R is a UI-ring. Then:

(a) Each VN-regular element of R is idempotent.

(b) The idempotent elements of R from a subring of the centre of R.

(c) If R is VN-regular, then R is a Boolean ring with identity element.

PROOF: (a) Suppose (5) holds. By Corollary 2.6 and Lemma 2.7, the idempotents ax and xa are equal. A routine computation shows that

$$(1 - ax + a)(1 - ax + x) = 1,$$

and it follows that $(1 - ax + x)$ and $(1 - ax + a)$ are invertible elements of R. Since R is a UR-ring,

$$ax = x = a.$$

Thus $a^2 = a$ and (a) is established.

(b) Since each idempotent of R is in its centre, the idempotents are closed under multiplication. Since by Corollary 2.6, $2x = 0$ for each $x \in R$, they are closed under addition as well. So (b) holds.

(c) follows immediately from (b), so Theorem 2.9 holds. □

As usual let $R[x]$ denote the ring of polynomials with coefficients in R.

COROLLARY 2.10: Any element of a UI-ring R that is algebraic over Z_2 is an idempotent.

PROOF: Suppose $a \in R$ is algebraic over Z_2 with minimum polynomial $m(x) = x^n + \alpha_1 x^{n-1} + \ldots + \alpha_n$ where $\alpha_i \in Z_2$ for $1 \leq i \leq n$ and $n \geq 2$. If $\alpha_n = 1$, then $a(a^{n-1} + \alpha_1 a^{n-2} + \ldots + \alpha_{n-1}) = 1$, so $a = 1$ is idempotent since 1 is the only invertible element of R. So assume $\alpha_n = 0$ and write $m(x) = x^k p(x)$ where $p(x) \in Z_2[x]$ satisfies $p(0) = 1$ and k is a positive integer. If $k > 1$, then $(ap(a))^k = 0$ contrary to the fact that UI-rings are reduced. So $k = 1$ and $m(x) = xp(x)$. Since $p(0) = 1$ and $n \geq 2$, we may write $p(x) + 1 = xq(x)$ for some $q(x) \in Z_2[x]$. It follows that $x^2 q(x) = m(x) + x$ and hence that $a^2 q(a) = m(a) + a = a$. But every VN-regular element of a UI-ring is idempotent, and the corollary holds. □

This section concludes with a characterization of Boolean rings with identity element, whose proof is particularly simple.

THEOREM 2.11: A UR-ring R is a Boolean ring with identity element if and only if $a^3b^3 = b^3$ implies $ab = b$ whenever $a,b \in R$.

PROOF: The necessity is clear, so we may assume R is a UR-ring with identity element 1. Suppose $b \in A(1 + a + a^2)$, or since UR-rings have characteristic 2,

$$a^2b = ab + b. \tag{6}$$

Then $a^3b = a^2b + ab = (ab + b) + ab = b$. Multiplying this latter equation on the right by b^2 yields $a^3b^3 = b^3$, so by assumption, $ab = b$ and hence $a^2b = 0$ by (6). By Lemma 2.7, since $a^2b^2 = 0$, it follows that $(ab)^2 = ab$ since R is reduced. Using (6) again, we conclude that $b = 0$ and hence $1 + a + a^2 = 1$. Thus $a = a^2$ and R is a Boolean ring. □

3. PROPERTIES OF SUR-RINGS

In this section we will describe a necessary condition for a ring with identity element to be an SUR-ring, and we give an example of a UI-ring that fails to satisfy it. Thus the hypothesis of commutativity in Cohn's result [2] that every commutative UI-ring is an SUR-ring is essential.

Recall that an ideal P of a ring R is said to be a *prime* (resp. *completely prime*) ideal if $aRb \subset P$ (resp. $ab \in P$) implies $a \in P$ or $b \in P$, and R is called a *prime ring* (resp. *domain* if {0} is a prime (resp. completely prime) ideal of R. There are many thorough discussions of the properties of prime ideals in the literature (e.g. the one given in [9], chapter 4) but no corresponding one seems to exist for completely prime ideals. What is known seems to be widely scattered in the literature. We cite a few results that are noteworthy or will be needed below.

A prime ideal P of R is completely prime if and only if R/P is reduced [5], Lemma 1.1.1. (7)

Every minimal prime ideal of a reduced ring is completely prime [7] (8)

A ring in which $A(a) = 0$ whenever $a \neq 0$ is called a *domain*.

Every reduced ring is a subdirect product of domains; that is, the intersection of the completely prime ideals of a reduced ring is $\{0\}$ [5], Theorem 1.1.1. (9)

While it is hard to believe that the following lemma is new, no reference for it comes to mind.

LEMMA 3.1: A prime ideal P of a ring R is completely prime if and only if $P \cap S$ is a completely prime ideal of S for every subring S of R.

PROOF: Suppose P is a completely prime ideal of R and S is a subring of R. If $a,b \in S$ and $aSb \subset P \cap S \subset P$, then $a \in P \cap S$ or $xb \in P \cap S$ for all $x \in S$. Either $S = P$ and $a \in P$ or there is an $x \in S \setminus P$ and $b \in P$ since P is completely prime. Hence $P \cap S$ is prime.

Suppose conversely that $P \cap S$ is a prime ideal of S for every subring S or R and suppose $ab \in P$ for some a,b in R. Let S' denote the subring of R generated by a and b. Then $aS'b \subset P \cap S'$, so $a \in P$ or $b \in P$. Thus Lemma 3.1 holds. □

The main positive result of this section follows.

THEOREM 3.2: If S is a SUR-ring with identity element 1, then:

Every $a \neq 1$ in S is contained in a proper completely prime ideal. (*)

The converse holds if S is commutative.

PROOF: Suppose R is a UR-ring with identity element 1 containing S and suppose $1 \neq a \in S$. Since $A_R(a) = \{x \in R: ax = 0\} \neq \{0\}$, a is contained in a minimal prime ideal P' of R as is shown in [12], Theorem 3.10. As was noted in (8) above, P' is a completely prime ideal of R, so $P = P' \cap S$ is a completely prime ideal of S by Lemma 3.1. Since $1 \in S \setminus P'$, the ideal $P' \cap S$ is proper and contains a. So (*) holds. □

Clearly every ring that satisfies (*) is a UI-ring, so by Cohn's result that every commutative UI-ring is an SUR-ring, the converse holds if (*) holds and S is commutative.

Next, we give an example of a UI-ring that is not a SUR-ring.

EXAMPLE 3.3: Let W denote the Weyl algebra over Z_2 in the noncommuting indeterminates x, y subject to the relation $yx = xy + 1$; see [3], section 0.8, or [8], Example 1. Clearly W is a UI-ring, but the element x is in no proper (completely prime) ideal, so W fails to be an SUR-ring by Theorem 3.2.

This note concludes with a number of problems and remarks.

(A) Is there a UR-ring that fails to be commutative? If it could be shown that every ring S that satisfies (*) is an SUR-ring, then many such UR-rings could be constructed.

(B) Suppose R is a UR-ring with centre Z(R). Must Z(R) be a UR-ring? That is, must $A(a) \cap Z(R)$ be nonzero if $1 \neq a \in R$? If R also satisfies a polynomial identity over Z(R), then the answer is yes since in a PI-ring R, every nonzero ideal has nonzero intersection with Z(R) by [11]; see also [1], p. 472.

Before stating our last problem, it should be noted that the literature does not abound with examples of reduced rings that contain prime ideals that fail to be completely prime. One such is the ring W of Example 3.3. It is noted in [8], Example 1, that $M = (x^2 + 1)W + (y^2 + 1)W$ is a maximal (and hence prime) ideal of W that is not completely prime.

Other such examples may be obtained as follows. Suppose R is a primitive ring that is a domain. (For example, let $R = Z_2\langle x,y\rangle$ denote the free ring with identity over Z_2 in noncommuting indeterminates x, y; see [6], section 4.1.) If $n > 1$ is a positive integer, since R fails to be isomorphic to the ring D_n of all $n \times n$ matrices for any with entries in some division ring D, there is a division ring D, subring S or R and a homomorphism ϕ of S onto D_n; see [4], Theorem 2.1.4. Then the kernel of ϕ is a maximal (and hence prime) ideal of S that fails to be completely prime. In particular $Z_2\langle x,y\rangle$ contains a subring S with a prime ideal that fails to be completely prime. This motivates our last problem.

(C) Is $Z_2\langle x,y\rangle$ and SUR-ring?

ACKNOWLEDGEMENTS: I am indebted to Professor John Clark for the reference to [8] and stimulating conversations. Most of this research was done while the author was a John H. Van Vleck Visiting Professor at Wesleyan University, Middletwon, Connecticutt, U.S.A.

REFERENCES

[1] E. Armendariz, A note on extensions of Baer and P.P.-rings. J. Austral. Math. Soc. 18 (1974), 470-473.

[2] P.M. Cohn, Rings of zero divisors, Proc. Amer. Math. Soc. 9 (1958), 909-914.

[3] P.M. Cohn, Free Rings and Their Relations, Academic Press, New York, 1971.

[4] I. Herstein, Noncommutative Rings, Mathematical Association of America, Providence, RI, 1968.

[5] I. Herstein, Rings with Involution, University of Chicago Press, Chicago 1976.

[6] N. Jacobson, Basic Algebra II, W.H. Freeman, San Francisco, 1980.

[7] K. Koh, On functional representation of a ring without nilpotent elements, Canad. Math. Bull 14 (1971), 349-352.

[8] M. Lorenz, Completely prime ideals in Ore extensions, Commun. Algebra 9 (1981), 1227-1232.

[9] N. McCoy, The Theory of Rings, Chelsea, New York, 1973.

[10] T. Porter, Cohn's rings of zero divisors, Arch. Math. 43 (1984), 340-343.

[11] L. Rowen, Some results on the center of a ring with polynomial identity, Bull. Amer. Math. Soc. 79 (1973), 219-223.

[12] N. Thakare and S. Nimbhorkar, Spaces of minimal prime ideals of a ring without nilpotent elements, J. Pure Appl. Algebra 27 (1983), 75-85.

M. Henriksen
Department of Mathematics
Harvey Mudd College
Claremont
CA 91711
U.S.A.

GERHARD HOFER

Reachability in machines

ABSTRACT: If the state set and the input set of an automaton are R-modules then near-rings are useful in the study of automata (see [3] or [9]). These near-rings, called syntactic near-rings, consist of mappings from the state set Q of the automaton into itself. The zerosymmetric part N_0 of syntactic near-rings (which is a commutative ring with identity) is suitable for determining reachability in automata. In this note we give a summary of results concerning reachability in machines.

A *semiautomaton* is a triple $A = (Q,A,F)$, consisting of two nonempty sets Q and A (the *state set* and the *input set*) and a function $F: Q \times A \to Q$, called the *next state function* of A (see [8] or [11] for the theory of these creatures). A is *finite* if Q is finite. If Q and A are R-modules and F is an R-homomorphism then we call A an R-module semiautomaton (MSA), briefly written as $A = (Q,A,F)_R$. For any semiautomaton $A = (Q,A,F)$ we get a collection of mappings $F_a: Q \to Q$, one for each $a \in A$, which are given by $F_a(q) := F(q,a)$. If the input a_1 is followed by the input a_2, the semiautomaton moves from the state $q \in Q$ first into $F_{a_1}(q)$ and then into $F_{a_2}(F_{a_1}(q))$. In the case of MSAs we are also able to study $F_{a_1} + F_{a_2}$ (defined pointwisely). Hence $\{F_a \mid a \in A\} \cup \{F_\Lambda\}$ (where Λ denotes the empty input) and all its sums and products (= composition of maps) are suitable to describe the answer of the automaton receiving superpositions of sequences of inputs (see [7]). Because of $F_a(q) = F(q,a) = F(q,0) + F(0,a) = F_0(q) + F_a(0)$ we get $F_a = F_0 + \bar{F}_a$, where F_0 is an R-homomorphism, while $\bar{F}_a$ is the map with constant value $F_a(0)$. Since $F_\Lambda = id_Q$ so $\{F_a \mid a \in A\} \cup \{F_\Lambda\}$ and all its sums and products are a subset of $M_{aff}(Q) = \{f: Q \to Q \mid f \text{ affine map}\}$ which is an abstract affine near-ring (see [12] for the theory of near-rings). The subnear-ring N(A) of $M_{aff}(Q)$ generated by id_Q and all F_a's ($a \in A$) is called the *syntactic near-ring* of A. Just the zerosymmetric part $N_0(A) := (N(A))_0$ (which is a ring) of the syntactic near-ring is suitable for studying reachability in machines (see [13]). It is proved that $N_0(A) = \{z_0 id_Q +$

$z_1F_0^1 + \dots + z_nF_0^n \mid z_i \in \mathbf{Z},\ n \in \mathbb{N}_0\} = \langle id_Q, F_0\rangle_{ri}$ (see [3]). Hence $N_0(A)$ is a commutative ring with identity. For the following, see [5]. A subset $U \subseteq Q$ is called *reachable from q* ($\in Q$) if for all $q' \in U$ there is a $n_0 \in N_0(A)$ such that $n_0(q) = q'$. One sees immediately that U is reachable from q iff $U \subseteq N_0(A)q$. If we consider Q as an $N_0(A)$-module via $n_0 \cdot q := n_0(q)$ then reachable sets can be found within $N_0 := N_0(A)$ itself by the homomorphism theorem. That means $N_0q = \mathcal{R}q$ where $\mathcal{R}$ is a system of representatives of the factor N_0-module $N_0/_{(0:q)}$ (this representation of N_0q is "as small as possible", which is important for the explicit calculation of reachable sets; see the examples below). Conversely, if L is an ideal of N_0 then $N_0\bar{q} = \mathcal{R}\bar{q}$ for all $\bar{q} \in \text{dep } L := \{q \in Q \mid L = (0:q)\}$ where $\mathcal{R}$ is a system of representatives of the factor N_0-module $N_0/_L$. In fact, the set dep L is "dependent from L" and can be empty if L is not the annihilator of one element. If dep $L \neq \emptyset$ then all $q \in \text{dep } L$ have the same reachability $\mathcal{R}q$ and one can determine $\mathcal{R}$ in some special cases, as will be shown in the following. Also we will give conditions which force dep L to be nonempty and conditions which allow one to compute dep L more explictly.

In the following $\langle X\rangle_{id}$ ($\langle X\rangle_{gp}$, $\langle X\rangle_{ri}$) indicates the ideal (group, ring) generated by X. $L \vartriangle N_0$ means $L \trianglelefteq N_0$ and $L \neq N_0$. $\mathcal{L}_L := \{J \trianglelefteq N_0 \mid L \subsetneqq J$ and J/L is a simple N_0-module$\}$ for a fixed $L \trianglelefteq N_0$.

<u>THEOREM 1</u>: Let $A = (Q,A,F)_R$ be an MSA and $L = \langle E_L\rangle_{id} \vartriangle N_0(A)$ (=: N_0). If N_0 possess a composition series then

$$\text{dep } L = \bigcap_{e_\ell \in E_L} \text{Ker } e_\ell \setminus \bigcup_{J \in \mathcal{L}_L} \Big(\bigcap_{e_j \in E_J} \text{Ker } e_j \Big).$$

With the notation of Theorem 1 (and since N_0 is generated by id_Q) we get the following:

<u>COROLLARY 1</u>: (a) If L is a maximal ideal then

$$\text{dep } L = \bigcap_{e_\ell \in E_L} \text{Ker } e_\ell \setminus \{0\}.$$

(b) If N_0 is a principal ideal ring then

$$\text{dep } L = \text{Ker } e_\ell \setminus \bigcup_{J \in L_L} \text{Ker } e_j$$

(where all ideals X are generated by e_X). If, moreover, L is a maximal ideal, then dep L = Ker $e_\ell \setminus \{0\}$.

If $(G,+)$ is an abelian group then $\mathbb{Z}_n g := \{0, g, 2g, \ldots, (n-1)g\}$ for $g \in G$, $n \in \mathbb{N}$. With this notation we can easily determine a system of representatives in the following:

<u>THEOREM 2</u>: Let $A = (Q,A,F)_R$ be an **M**SA where $N_0 := N_0(A) = \sum_{i=0}^{m-1} \langle F_0^i \rangle_{gp}$ has finite characteristic. Let $q \in Q$ and

$$a_i := \min\{n \in \mathbb{N} \mid nF_0^i(q) \in \sum_{j=0}^{i-1} \langle F_0^j(q) \rangle_{gp}\}, \quad i \in \{1,\ldots,m-1\},$$

$$a_0 := \text{ord } q,$$

then the following hold:

(a) $R = \mathbb{Z}_{a_0} \text{id} + \mathbb{Z}_{a_1} F_0 + \ldots + \mathbb{Z}_{a_{m-1}} F_0^{m-1}$ is a system of representatives of $N_0/(0:q)$;

(b) $N_0 q = \mathbb{Z}_{a_0} q + \mathbb{Z}_{a_1} F_0(q) + \ldots + \mathbb{Z}_{a_{m-1}} F_0^{m-1}(q)$.

It remains to clear in which cases dep L is forced to be nonempty (see [5] and [10]).

<u>THEOREM 3</u>: Let $A = (Q,A,F)_R$ be an MSA. If Q is a free $N_0(A)$-module, then the following holds:

$N_0(A)$ is an artinian principal ideal ring $\Rightarrow$ dep L $\neq \emptyset$ for all ideals L of $N_0(A)$.

In fact, it is rarely the case that Q itself is reachable from just one $q \in Q$. The question arises how many states q are necessary to reach Q (see

[6]). This motivates the following:

<u>DEFINITION 1</u>: Let $A = (Q,A,F)_R$ be an MSA and U a subset of Q. U is called k-*reachable* : $\Longleftrightarrow \exists\, q_1,\dots,q_k \in Q\ \forall q \in U\ \exists\, n_1,\dots,n_k \in N_0(A) : n_1(q_1) + \dots + n_k(q_k) = q$. A is called k-*reachable* if Q is k-reachable.

With $N_0 := N_0(A)$ it holds again that U is k-reachable $\Longleftrightarrow \exists\, q_1,\dots,q_k \in Q$: $U \subseteq N_0q_1 + \dots + N_0q_k$. Since N_0 is a ring with identity, hence Q is $|Q|$-reachable (where $|Q|$ denotes the cardinality of Q) because $Q = \sum_{q \in Q} N_0 q$. It is interesting in which cases the sum is direct and to find k-reachability where k is as small as possible.

<u>DEFINITION 2</u>: Let $A = (Q,A,F)_R$ be an MSA and U a subset of Q. U is called *strictly* k-*reachable* if U is k-reachable and

$$U = \sum_{i=1}^{k}{}^{\bullet}\ N_0 q_i.$$

A is called *strictly* k-*reachable* if Q is strictly k-reachable.

<u>THEOREM 4</u>: Let $A = (Q,A,F)_R$ be a finite MSA and U a $N_0(A)$-submodule of Q. Then the following hold:

(a) If $N_0(A)$ is semisimple then U is strictly k-reachable for some $k \in \mathbb{N}$.

(b) If U is a free $N_0(A)$-module then U is strictly k-reachable and $k = \dim U$.

If U is strictly k-reachable then

$$U = \sum_{i=1}^{k}{}^{\bullet}\ N_0 q_i = \sum_{i=1}^{k}{}^{\bullet}\ R_i q_i$$

where R_i is a system of representatives of $N_0/_{(0:q_i)}$, $i \in \{1,\dots,k\}$. If U is free then $N_0 = R_i$ for all $i \in \{1,\dots,k\}$. Otherwise all R_i's can be explictly found using Theorem 2. For calculating examples we shall need a few more previous remarks. For the following, see [4].

Let R be a ring with identity 1, char R = n $\in$ **N**. If R = $\langle 1,r\rangle_{gp}$ where $r^2 = \alpha 1 + \beta r$ for some $\alpha,\beta \in \mathbf{Z}$, ord r = n and $\langle 1\rangle \cap \langle r\rangle = \{0\}$ then R is a *syntactic ring* (s.r.) with *generator* r and *syntactic triple* (s.tr.) $(\alpha,\beta;n) \in \mathbf{Z} \times \mathbf{Z} \times \mathbf{N}$. We abbreviate this situation by $R(\alpha,\beta;n)$. $R(\alpha,\beta;n) = \langle 1\rangle_{gp} \dotplus \langle r\rangle_{gp}$ has cardinality n^2. The map $h:z_0 1 + z_1 r \to (z_0,z_1)$ is a ring isomorphism between $(R,+,\cdot)$ and $(\mathbf{Z}_n \times \mathbf{Z}_n,+,\bar{\cdot})$ where the addition is componentwise and $\bar{\cdot}$ means the "syntactic multiplication". That means $(x_0,x_1) \bar{\cdot} (y_0,y_1) = (x_0y_0 + x_1y_1\alpha, x_0y_1 + x_1y_0 + x_1y_1\beta)$. $(\mathbf{Z}_n \times \mathbf{Z}_n,+,\bar{\cdot})$ is a s.r. itself with generator (0,1), identity (1,0) and s.tr. $(\alpha,\beta;n)$. If $(k,r,s;a) \in \mathbf{N} \times \mathbf{N} \times \mathbf{N} \times \mathbf{N}$ is a *divisor quadruple* (d.q) of n (that means k,r,s,kr,ks divide n and $a \in \{1,\ldots,k\}$ with gcd(k,a) = 1) then H(k,r,s;a) denotes a *Remak subgroup*. If (k,r,s;a) is a *positive divisor quadruple* (p.d.q) (that means a d.q. where $s \leq r$) then $H(k,r,s;a) = \langle r,sa\rangle_{gp} \dotplus \langle kr,0\rangle_{gp}$ is called a *positive Remak subgroup* (pRs). If (k,r,s;a) is a *negative divisor quadruple* (n.d.q.) (that means a d.q. where $r < s$) then $H(k,r,s;a) = \langle r,sa\rangle_{gp} \dotplus \langle 0,ks\rangle_{gp}$ is a *negative Remak subgroup* (nRs). All subgroups of $\mathbf{Z}_n \times \mathbf{Z}_n$ are nRs or pRs and H(k,r,s;a) has cardinality n^2/krs. nRs are never ideals of $(\mathbf{Z}_n \times \mathbf{Z}_n,+,\bar{\cdot})$. But pRs's H(k,r,s;a) are ideals of $(\mathbf{Z}_n \times \mathbf{Z}_n,+,\bar{\cdot})$, $n = p^t$, where p is prime and t $\in$ **N**, iff $k\frac{r}{s}$ divides $r^2/s^2 + a\beta\frac{r}{s} - a^2\alpha$. Let $(\mathbf{Z}_p \times \mathbf{Z}_p,+,\bar{\cdot})$ be a s.r. where p is prime, then $(\mathbf{Z}_p \times \mathbf{Z}_p,+,\bar{\cdot})$ is a principal ideal ring. If $(\mathbf{Z}_{p^2} \times \mathbf{Z}_{p^2},+,\bar{\cdot})$ is a s.r. with s.tr. $(\alpha,\beta;p^2)$, where p is prime, and $\alpha \not\equiv 0 \pmod{p^2}$, $1 + a\beta \not\equiv a^2\alpha \pmod{p^2}$ for all $a \in \{1,\ldots,p-1\}$, then $(\mathbf{Z}_{p^2} \times \mathbf{Z}_{p^2},+,\bar{\cdot})$ is a principal ideal ring, too.

Let $n = \prod_{i\in A} p_i^{t_i}$ and $R(\alpha,\beta;n)$ be a s.r. Then $R(\alpha,\beta;n)$ is ring isomorphic to $\bigoplus_{i\in A} R_i(\alpha,\beta;p_i^{t_i})$ by the map $(z_0 + z_1r) \to ((z_0 + z_1r),\ldots,(z_0 + z_1r))$. So $\bigoplus_{i\in A} T_i$ is an ideal of $\bigoplus_{i\in A} R_i$ iff T_i is an ideal of R_i, for all $i \in A$. All ideals are given by $\bigoplus_{i\in A} T_i$, since all $(R_i,+)$ are p_i-groups. If all R_i are principal ideal rings then $\bigoplus_{i\in A} R_i$ is, too. Together with the information above we are able to determine the ideals of all s.r. $R(\alpha,\beta;n)$, n $\in$ **N**, and get a lot more principal ideal rings (see [4] for a detailed prescription), hence a larger number of examples for the following considerations.

Let $A = (Q,A,F)_R$ be an MSA where $(Q,+) = ((\mathbf{Z}_n)^m,+)$, n,m $\in$ **N**. Then F_0 corresponds to a matrix $G = (g_{ij}) \in (\mathbf{Z}_n)^m_m$ (= the collection of all m $\times$ m matrices over $\mathbf{Z}_n$). If at least one g_{ij} $(i \neq j)$ is invertible then

$N_0(A) = \langle I\rangle_{gp} \dotplus \langle G\rangle_{gp} + \langle G^2\rangle_{gp} + \dots + \langle G^{m-1}\rangle_{gp}$ (because of the theorem of Cayley-Hamilton and since $N_0(A) = \langle I,G\rangle_{ri}$). If $m = 2$ then $N_0(A)$ is forced to be a syntactic ring (with generator G and s.tr. $(\alpha,\beta;n)$ for certain $\alpha,\beta \in \mathbf{Z}$). Now we have the ingredients for the following:

<u>EXAMPLE</u>: Let $A = (Q,A,F)_R$ be an MSA.

(a) Let $Q = \mathbf{Z}_4 \times \mathbf{Z}_4$ and F_0 correspond to $G = \begin{pmatrix} 1 & 1 \\ 0 & 2 \end{pmatrix}$ $(\in (\mathbf{Z}_4)_2^2)$. Then $G^2 = 2I + 3G$, hence $N_0(A)$ is a s.r. with s.tr. $(\alpha,\beta;n) = (2,3;4)$. Since $\alpha \neq 0 \pmod 4$ and $1 + \beta \neq \alpha \pmod 4$, hence $N_0(A)$ is a principal ideal ring. Since $b = (0,1)$, $Gb = (1,2)$ are linearly independent, hence Q is a one-dimensional $N_0(A)$-module (that means all dep L are nonempty, L an ideal of $N_0(A)$). Consider $L = \langle 2I + 2G\rangle_{gp} \dotplus \langle 2I\rangle_{gp}$. Then $h(L) = \langle 2,2\rangle_{gp} \dotplus \langle 2,0\rangle_{gp} = H(1,2,2;1) = H(k,r,s;a)$ is an ideal of $(\mathbf{Z}_4 \times \mathbf{Z}_4,+,\bar{\cdot})$ since $k\ r/s$ divides $r^2/s^2 + \beta\ r/s - \alpha$. Hence L is an ideal (with four elements) of $N_0(A)$, generated by one element, namely $L = \langle 2I\rangle_{id}$. The only three subgroups of $\mathbf{Z}_4 \times \mathbf{Z}_4$ with eight elements are $H(2,1,1;1)$, $H(1,2,1;1)$ and $H(1,1,2;1)$. The last one is a nRs, hence not an ideal. The same argument as above yields that $H(2,1,1;1) = \langle 1,1\rangle_{gp} \dotplus \langle 2,0\rangle_{gp}$ and $H(1,2,1;1) = \langle 2,1\rangle_{gp} \dotplus \langle 2,0\rangle_{gp}$ are ideals. They are generated by $(1,1)$, $(0,1)$, respectively. Thus $\langle I + G\rangle_{id}$, $\langle G\rangle_{id}$ are the only two ideals of $(\mathbf{Z}_4 \times \mathbf{Z}_4,+,\bar{\cdot})$ with eight elements. Hence $\mathcal{L}_L = \{\langle G\rangle_{id}, \langle I + G\rangle_{id}\}$. Since Ker $G = \{(0,0), (2,2)\}$, Ker $I + G = \{(0,0), (2,0)\}$, Ker $2I = \{(0,0), (2,0), (0,2), (2,2)\}$, hence dep L = Ker $2I \setminus$ Ker $G \cup$ Ker $I + G = \{(0,2)\}$. Since $\text{ord}(0,2) = \text{ord}\ G(0,2) = 2$, hence $R = \mathbf{Z}_2 I + \mathbf{Z}_2 G$ is a system of representatives of $N_0/_L$ and $N_0(A)q = \mathbf{Z}_2 q + \mathbf{Z}_2 Gq$, $q \in$ dep L, holds.

(b) Let $(Q,+) = ((\mathbf{Z}_n)^4,+)$, $n \in \mathbf{N}$, and F_0 correspond to $G = \begin{pmatrix} \bar{G} & 0 \\ 0 & \bar{G} \end{pmatrix}$ $(\in (\mathbf{Z}_n)_4^4)$ and $\bar{G} = \begin{pmatrix} 1 & 1 \\ 0 & 1 \end{pmatrix}$ $(\in (\mathbf{Z}_n)_2^2)$. Then $G^2 = I - 2G$ hence $N_0 := N_0(A)$ is a s.r. with s.tr. $(1,n-2;n)$. We consider $B = \{b_1,b_2\} = \{(0,1,0,0), (0,0,0,1)\}$. $Gb_1 = (1,1,0,0)$, $Gb_2 = (0,0,1,1)$ hence b_1,b_2,Gb_1,Gb_2 are linearly independent. Thus Q is a two-dimensional N_0-module. Thus $Q = N_0 b_1 \dotplus N_0 b_2$ that means A is strictly 2-reachable. If $n = p$ is prime then N_0 is an artinian principal ideal ring, hence dep L is nonempty for all ideals L of N_0. If $n = p^2$, p prime, then, in some cases (dependent from α, β), N_0 is a principal ideal ring, too (compare the lines before the example. If $n = \prod_{i \in A} p_i^{t_i}$, then (because of our previous remarks) we have just to investigate all s.r. $(\mathbf{Z}_{p_i^{t_i}},+,\cdot)$, $i \in A$, for a decision if N_0 is a principal ideal ring which can be one by methods given in [4].

(c) Let $(Q,+) = ((\mathbf{Z}_n)^3,+)$, $n \in \mathbf{N}$, and F_0 correspond to $G = \begin{pmatrix} 1 & 0 & 0 \\ 0 & 0 & 0 \\ 0 & 1 & 1 \end{pmatrix} (\in (\mathbf{Z}_n)^3_3)$. Then $G^2 = G$ hence $N_0 := N_0(A)$ is a s.r. with s.tr. $(\alpha,\beta;n) = (0,1;n)$. If $n = p$ is prime then N_0 is isomorphic (by h as above) to $(\mathbf{Z}_p \times \mathbf{Z}_p,+,\cdot)$, hence N_0 is a principal ideal ring. $H(1,1,1;1) = \mathbf{Z}_p \times \mathbf{Z}_p$ and $H(1,p,p;1) = \{(0,0)\}$. $H(p,1,1;a)$'s, $a \in \{1,\dots,p-1\}$, are ideals iff p divides $1 + a\beta - a^2\alpha$. Since $(\alpha,\beta) = (0,1)$, hence just $H(p,1,1;p-1) = \langle 1,p-1\rangle_{gp}$ is an ideal. That $H(1,p,1;1) = \langle 0,1\rangle_{gp}$ is an ideal follows analogously. $H(1,1,p;1)$ is a nRs, hence never an ideal. Now we have already exhausted all possibilities for Rs of $Z_p \times Z_p$. Thus $\langle G\rangle_{gp}$ and $\langle I + (p-1)G\rangle_{gp}$ are the only nontrivial ideals of N_0 (with p elements). Their intersection is trivial, hence N_0 is semisimple.

$$\text{Ker } G = \{(0,x,-x) \mid x \in \mathbf{Z}_p\}, \text{ Ker } I + (p-1)G = \{(x,0,z) \mid x,z \in \mathbf{Z}_p\}.$$

$\text{dep}\langle G\rangle_{gp} = \text{Ker } G \setminus \{0\}$, $\text{dep}\langle I + (p-1)G\rangle_{gp} = \text{Ker } I + (p-1)G \setminus \{0\}$, $\text{dep } \{0\} = Q \setminus \text{Ker } G \cup \text{Ker } I + (p-1)G$. $q_1 := (0,1,0)$ ($\in \text{dep } \{0\}$) yields (the "largest possible" system of representatives) $R_1 := \mathbf{Z}_p I + \mathbf{Z}_p G$ and $N_0 q_1 = R_1 q_1 = \{0\} \times \mathbf{Z}_p \times \mathbf{Z}_p$. (For example) $q_2 := (1,0,0)$ ($\in \text{dep } I + (p-1)G$) yields $R_2 = \mathbf{Z}_p I$ and $N_0 q_2 = R_2 q_2 = \mathbf{Z}_p \times \{0\} \times \{0\}$. All together we get $Q = R_1 q_1 \dotplus R_2 q_2$, which means A is strictly 2-reachable.

Now we give a method for the explicit determination of reachability in A, if $A = (Q,A,F)_R$ is a finite MSA where $N_0 := N_0(A)$ is a syntactic ring and U ($\subseteq Q$) is an N_0-module:

(1) Determine if Q is a free N_0-module and if N_0 is a (semisimple) principal ideal ring.
(2) Determine all ideals L of N_0 and dep L.
(3) Determine a system of representatives R of $N_0/_L$ for all ideals L of N_0.
(4) Rq is reachable from q for all $q \in$ dep L.
(5) Form $U = R_1 q_1 \dotplus \dots \dotplus R_k q_k$ where the q_i's belong to some favourable dep L and the R_i's are systems of representatives.

Finally we note that the "R-module structure" of Q in $A = (Q,A,F)_R$ was not necessary for our considerations. In fact, if Q is just an abelian group and F a group homomorphism (more generally a universal algebra (Q,Ω) which is an abelian Ω-group and F is an Ω-homomorphism, see [2]) then all statements above hold as well (see [5]). If Q is a (not necessarily) abelian group (Ω-group) then $N_0(A)$ is a near-ring (a subnear-ring of Aff(Q) = the near-ring generated by all Ω-homomorphisms and constant maps from Q to Q, see [1]) and Q can be considered as $N_0(A)$-group (or $N_0(A)$-near-module) via $n_0 \cdot q := n_0(q)$. Then left ideals of $N_0(A)$ play the same role as ideals in the abelian case. $N_0(A)$-module Q has to be replaced by $N_0(A)$-group Q and so on. Theorem 1 and Theorem 3 can then be generalized (see [5]). But there seems no hope to find explicit expressions for a system of representatives of the factor $N_0(A)$-group $N_0(A)/_{(0:q)}$ as in Theorem 2. And no hope to find explicitly all left ideals of $N_0(A)$ like the method given [4].

REFERENCES

[1] Feigelstock, S., The near-ring of generalized affine transformations, Bull. Austral. Math. Soc. 32 (1985), 345-349.

[2] Grätzer, G., Universal Algebra, 2nd edition, Springer, Berlin, 1979.

[3] Hofer, G., Near-rings and group automata, Doctoral dissertation, Univ. Linz, 1986.

[4] Hofer, G., Syntactic rings, Institutsbericht No. 356, Math. Inst., Univ. Linz, 1987.

[5] Hofer, G., Left ideals and reachability in machines, Institutsbericht No. 360, Math. Inst., Univ. Linz, 1987.

[6] Hofer, G., k-Reachability, Institutsbericht No. 362, Math. Inst., Univ. Linz, 1987.

[7] Hofer, G. and Pilz, G., Group automata and near-rings, Contrib. Gen. Algebra 2, (1983), 153-162.

[8] Holcombe, M., Algebraic Automata Theory, Cambridge University Press, Cambridge, 1982.

[9] Holcombe, M., The syntactic near-ring of a linear sequential machine, Proc. Edinb. Math. Soc. 26 (1983), 15-24.

[10] Jaegermann, M. and Krempa, J., Rings in which ideals are annihilators, Fundamenta Math. 76 (1972), 95-107.

[11] Lidl, R. and Pilz, G., Applied Abstract Algebra, Springer, New York, 1984.

[12] Pilz, G., Near-Rings, 2nd edition, North-Holland, Amsterdam, 1983.

[13] Pilz, G., Strictly connected group automata, Proc. R. Irish Acad. 86A, No. 2 (1986), 115-118.

G. Hofer
Institut für Mathematik
Johannes Kepler Universität
Linz
A-4040 Linz
Austria

PAUL E. JAMBOR

The radical of splitting ring extensions

The objective of this paper is to lay elementary foundations to the study of one-sided splitting ring extensions, which appear to be natural generalizations of classical Everett extensions, semitrivial extensions [8], monoid rings and skew-polynomial rings.

All rings considered are associative unless specified otherwise and in what follows R and S stand for rings with identity $1 \neq 0$.

The most general form of a right splitting extension is given by a ring monomorphism $i:R \to S$ preserving the identity 1 and $p \in \mathrm{Hom}_R(S_R,R_R)$ such that $p(1) = 1$. Here, the right R-module structure of S_R is given by $s\cdot r = si(r)$. Without loss of generality we may redefine the extension by requiring that i is a subring inclusion. Then p is an epimorphism and $K_R = \mathrm{kernel}(p)$ is a direct summand of S_R.

The class of all splitting extensions of R is closed under compositions (i.e. if $R_1 \subseteq R_2 \subseteq R_3$ is a chain of extensions and $p_1:R_2 \to R_1$ and $p_2:R_3 \to R_2$ are the corresponding epimorphisms, then $R_1 \subseteq R_3$ with the epimorphism p_1p_2 is again a splitting extension of R_1). Notice that by using compositions we can build up generalized triangular and full matrix rings [6], [8].

Given a ring automorphism σ of R and a σ-derivation δ (δ is abelian group endomorphism of R such that $(rs)^\delta = r^\delta s^\sigma + rs^\delta$, for $r,s \in R$) we denote the general skew-polynomial ring over R by $R[x;\sigma,\delta]$, where the commutation is subject to $rx = xr^\sigma + r^\delta$ ([4], p. 34). Define

$$R[x;\sigma] = R[x;\sigma,0] \quad \text{and} \quad R[x;\delta] = R[x;1,\delta].$$

Since every field extension, or more generally, an extension of an artinian semisimple ring, is splitting, a meaningful complete classification of splitting extensions of a given ring does not seem to be feasible. (An extension S of a commutative ring R such that S_R is projective is also splitting [3].) However, by imposing conditions which emulate particular classes of splitting extensions, e.g. monoid rings or matrices, we can obtain results valid for families of ring extensions larger than those we wanted to emulate. Advantage is a unified treatment allowing us to see interdependence of particular classes of splitting extensions.

For the used definitions and notation the reader is referred to [1] or [5]. Let us recall some of the less frequently used terminology. Let T be also a ring and ${}_TA_R$ and ${}_SB_R$ be bimodules. The set of all homomorphisms $\mathrm{Hom}_R({}_TA_R, {}_SB_R)$ is equipped with an S-T bimodule structure given by $(sft)(a) = s(f(ta)$, for $s \in S$, $t \in T$ and $a \in A$. Consequently, when we consider right homomorphisms we apply the arguments to the right of the homomorphism f and the composition of two homomorphisms f, g is given by $(fg)(a) = f(g(a))$. For the left homomorphisms, change it *mutatis mutandis*.

A subset L of a ring is said to be right (left) T-nilpotent if for every sequence $a_1, a_2, \ldots$ in L there is an n such that $a_n \ldots a_2 a_1 = 0$ $(a_1 a_2 \ldots a_n = 0)$.

1. STRUCTURE OF SPLITTING EXTENSIONS

THEOREM 1.1: Let ${}_RA_R$ be a bimodule endowed with a binary operation making it a ring (possibly non-associative and without identity), $\alpha \in \mathrm{Hom}_R(R \otimes_Z A_R, R_R)$, $\beta \in \mathrm{Hom}_R(A \otimes_R A_R, R_R)$ be such that

(i) $(rs \otimes a)^\alpha - (r \otimes sa)^\alpha = r(s \otimes a)^\alpha$,

(ii) $r(ab) - (ra)b = (r \otimes a)^\alpha b$.

(iii) $a(br) = (ab)r$,

(iv) $r(a \otimes b)^\beta - (ra \otimes b)^\beta = ((r \otimes a)^\alpha \otimes b)^\alpha - (r \otimes ab)^\alpha$,

(v) $(ar)b - a(rb) = a(r \otimes b)^\alpha$,

(vi) $a(bc) - (ab)c = (a \otimes b)^\beta c - a(b \otimes c)^\beta$,

(vii) $(a \otimes bc)^\beta - (ab \otimes c)^\beta = ((a \otimes b)^\beta \otimes c))^\alpha$,

for every choice of $r,s \in R$ and $a,b,c \in A$. Then $S = R \times A$ as the abelian group with multiplication given by

$$(r,a)(s,b) = (rs + (r \otimes b)^\alpha + (a \otimes b)^\beta,\ \ rb + as + ab)$$

is a right splitting extension of R, and it will be denoted by $R\ {}^\alpha\nabla^\beta\ A$ (tensor representation of S). Conversely, every right splitting extension of R arises in this way, up to a ring-isomorphism.

PROOF: Let $S = R \times A$ be the abelian group with the described multiplication. Then the distributivity of tensor product implies the left and right distributivity of the multiplication, and (1,0), obviously, serves as the identity. It takes a tedious checking to verify that the conditions (i)-(vii) and $(1 \otimes a)^{\alpha} = 0$, for every $a \in A$, are equivalent to the associativity of the multiplication. However, $(1 \otimes a)^{\alpha} = 0$, for every $a \in A$, is a direct consequence of the condition (i). Now, the projection map $p:S \to R$ is clearly a right R-homomorphism and $p(1,0) = 1$. Hence S with the defined multiplication is a right splitting extension.

Conversely, if $R \subseteq S$ and $p:S_R \to R_R$ is a right splitting extension then we can define $A_R = \ker(p)$, $(r \otimes a)^{\alpha} = p(ra)$, $(a \otimes b)^{\beta} = p(ab)$, $a.b = ab - p(ab)$, and the left R-module structure of A by $r.a = a - p(ra)$, for every $r \in R$ and $a,b \in A$. Since p is an epimorphism and R_R is projective, A_R is a direct summand and $S = R_R \oplus A_R$. Also, the associativity of S implies the conditions (i) through (vii).

Obviously, the correspondence between right splitting extensions of R and their tensor representations is 1-1 up to a ring-isomorphism which is stable on R. □

EXAMPLE 1.2: Let R be a ring with a derivation D (with respect to the identity automorphism) such that $D^2 = 0$ and the ideal $2(R)^D R \neq R$. Put $T = R/(2(R)^D R)$ and define $S = T \; {}^{\alpha}\nabla^{\beta} \; A$, where $\beta = 0$, ${}_TA_T = {}_TT_T$ with the multiplication given by $t*s = (t^{\delta})s$, and $(t \otimes a)^{\alpha} = (t^{\delta})a$ (δ is the derivation of T induced by D).

Then S is a right splitting extension of T isomorphic to a factor ring of $R[x;D]/(x^2)$.

Notice that $SA = (T^{\delta})T \oplus A$ is nilpotent if and only if T^{δ} is so (cf. Theorem 1.4).

The interdependence between splitting extensions and their tensor representations is illustrated in the following proposition which is left to the reader to verify. Notice that $\alpha = \beta = 0$ corresponds to classical Everett extensions and $\alpha = 0$ together with A being the zero-ring corresponds to semitrivial extensions [8].

PROPOSITION 1.3: Let $R \subseteq S$ and $p:S_R \to R_R$ be a right splitting extension and $S = R \;{}^{\alpha}\nabla^{\beta}\; A$ be its tensor representation. Then:

(i) The following are equivalent:

(a) $\alpha = 0$,

(b) $\alpha \in \mathrm{Hom}_R(R \otimes_R A_R, R_R)$,

(c) $p \in \mathrm{Hom}_R({}_RS, {}_RR)$,

(d) ${}_R(\ker(p)) \subseteq {}_RS$.

(ii) The following are equivalent:

(a) $\alpha = \beta = 0$,

(b) $\ker(p)$ is a left ideal of S,

(c) $\ker(p)$ is an ideal of S,

(d) p is a ring-epimorphism.

(iii) $\beta = 0$ if and only if $\ker(p)$ is a right ideal of S.

THEOREM 1.4: Let $R \subseteq S$ and $p:S_R \to R_R$ be a right splitting extension, $K = \ker(p)$ and $I = SK \cap R$. Suppose that M_S is an S-module and N_R is an R-module. Then the following assertions hold:

(i) If M_S is projective then $(M/MK)_R$ is (R/I)-projective. Conversely, if N_R is projective then $(N \otimes {}_RS)_S$ is projective.

(ii) If SK is right T-nilpotent and P_R is a projective cover of $(M/MK)_R$ then $(P \otimes {}_RS)_S$ is a projective cover of M_S.

(iii) Suppose that SK is right T-nilpotent and $(R/I)_R$ is projective. Then, M_S is projective if and only if $(M/MK)_R$ is projective and $M_S \simeq (M/MK) \otimes {}_RS_S$. In such a case, $(MK)_R$ is a direct summand of M_R.

(iv) If M_S is injective then $(M:K) = \{m \in M;\ m(SK) = 0\}$ is (R/I)-injective. Conversely, if N_R is injective then $(\mathrm{Hom}_R({}_SS_R, N_R))_S$ is injective.

(v) If SK is left T-nilpotent and E_R is an injective hull of $(M:K)_R$ then $(\mathrm{Hom}_R({}_SS_R, E_R))_S$ is an injective hull of M_S.

(vi) Suppose that SK is left T-nilpotent and ${}_R(R/I)$ is flat. Then M_S is injective if and only if $(M:K)_R$ is injective and $M_S \simeq (\mathrm{Hom}_R({}_SS_R, (M:K)_R))_S$. In such a case $(M:K)_R$ is a direct summand of M_R.

<u>PROOF</u>: (i) Suppose that M_S is projective. Without loss of generality we may assume that M_S is a direct summand of $(S^{(\omega)})_S = M_S \oplus N_S$, for some cardinal ω. Then $(SK)^{(\omega)} = (S^{(\omega)})(SK) = M(SK) \oplus N(SK) = MK \oplus NK$ and $(R/I)^{(\omega)} \simeq (S/SK)^{(\omega)} \simeq S^{(\omega)}/(S^{(\omega)}(SK)) \simeq M/MK \oplus N/NK$, and consequently M/MK is (R/I)-projective. (Take into account the fact that $SK = I \oplus K$ and SK is an ideal of S.) The converse statement follows from the Hom-tensor product adjoint duality ([5], p. 430).

(ii) Consider the following commutative diagram:

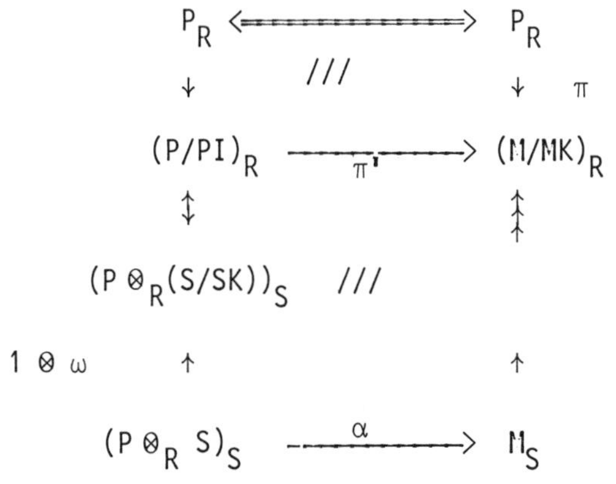

$$\begin{array}{ccccc}
 & P_R & \Longleftrightarrow & P_R & \\
 & \downarrow & /// & \downarrow & \pi \\
 & (P/PI)_R & \xrightarrow{\ \pi'\ } & (M/MK)_R & \\
 & \updownarrow & & \twoheaduparrow & \\
 & (P \otimes_R (S/SK))_S & /// & & \\
1 \otimes \omega & \uparrow & & \uparrow & \\
 & (P \otimes_R S)_S & \xrightarrow{\ \alpha\ } & M_S &
\end{array}$$

where π is the assumed projective cover; π' is the induced epic following from the fact that $SK = I \oplus K$, and that in turn implies $PI \subseteq \ker(\pi)$; ω is the projection ${}_RS \to {}_R(S/SK)$; $\leftrightarrow$ is the isomorphism induced by $S/SK \simeq R/I$; and $\twoheadrightarrow$ denotes natural projections.

Since both (P/PI) and (M/MK) have the trivial structure of right S-modules and in that structure π' is an S-homomorphism, the existence of α now follows from the projectivity of $P \otimes {}_R S_S$. Furthermore, since π is a superfluous epic, π' is a superfluous epic, too, and SK being right T-nilpotent implies that $((P \otimes {}_R S)SK)_S = \ker (1 \otimes \omega)_S$ is small ([1], p. 314). Hence the composition $\pi' (\leftrightarrow) (1_p \otimes \omega)$ is a superfluous epic. Similarly, the natural projection $M_S \to (M/MK)_S$ has the small kernel M(SK), and thus it is a superfluous epic. Therefore, α must be a superfluous epic, too ([1], 5.15, p.74).

(iii) Suppose that M_S is projective. Then, by using (i), $(M/MK)_R$ is (R/I)-projective, and since $(R/I)_R$ is projective, the hom-tensor product adjoint duality yields

$$\mathrm{Hom}_R((M/MK)_R,(\cdot)_R) \simeq \mathrm{Hom}_R(((M/MK) \otimes {}_{(R/I)}(R/I))_R,(\cdot)_R)$$

$$\simeq \mathrm{Hom}_R((M/MK)_R, \mathrm{Hom}_R({}_{(R/I)}(R/I)_R,(\cdot)_R)).$$

Hence $(M/MK)_R$ is projective. In particular, $(MK)_R$ is a direct command of M_R. Consider the map $g:((M/MK) \otimes {}_R S)_S \to (M/MK)_S$ given by $g(m\hat{} \otimes s) = (ms)\hat{}$, where m stands for the equivalence class of $m \in M$ modulo MK. Obviously, g is a well-defined S-epimorphism and since $((M/MK) \otimes {}_R S)_S$ is projective there exists $g' \in \mathrm{Hom}_S((M/MK) \otimes {}_R S)_S, M_S)$ such that $\tau g' = g$, where $\tau: M_S \to (M/M(SK))_S$ is the projection. Furthermore, SK being right T-nilpotent implies that $(MK)_S = (M(SK))_S$ is small in M_S and therefore g' is an epimorphism. Let $\Sigma\ ((m\hat{})_i \otimes s_i) \in \ker(g)$, where $s_i = r_i + k_i$, $r_i \in R$ and $k_i \in K$. Then $\Sigma((m\hat{})_i \otimes s_i) = ((\Sigma(m\hat{})_i r_i) \otimes 1) + \Sigma((m\hat{})_i \otimes k_i)$ and since $\Sigma((m\hat{})_i \otimes k_i) \in \ker(g)$ we obtain $(\Sigma(m\hat{})_i r_i) = 0$.

Thus $\ker(g) = ((M/MK) \otimes {}_R S)(SK)$ and that, thanks to SK being right T-nilpotent, implies $\ker(g') \subseteq \ker(g)$ is small, too. Hence g' is a projective cover and since M_S is projective, $M_S \simeq ((M/MK) \otimes {}_R S)_S$.

The converse statement follows directly from (i).

(iv), (v) and (vi) are dual statements to (i), (ii) and (iii) respectively, and the proofs can run along the same lines as above with slight modifications. □

EXAMPLE 1.5: Let R be a skew-field and ${}_R A_R$ be the set of all the countably infinite square upper triangular matrices over R with zeros on the main diagonal and only finitely many nonzero entries off the diagonal. Put $\alpha = \beta = 0$. Then A is a right T-nilpotent ideal of $S = R\ {}^{\alpha}\nabla^{\beta}\ A$, $J(S) = A$ (cf. Theorem 2.3) and S-projectives are free.

2. JACOBSON RADICAL

The Jacobson radical $J_R(M_R)$ of the right R-module M_R is the intersection of all maximal submodules of M_R. Precisely, $J_R(M_R) = \cap \ker(f)$, $f \in \mathrm{Hom}_R(M_R,T_R)$, where the intersection runs through all choices of f and simple modules T_R. Denote $J_R(R_R) = J(R)$ which can be characterized as the largest right (left) ideal consisting of right (left) quasi-invertible elements ([7], p. 196). In the following, $R \subseteq S$ and $p \in \mathrm{Hom}_R(S_R,R_R)$ is a given right splitting extension, $K_R = \ker(p)$ and $W = \{r \in R;\ Kr \subseteq J(S)\}$.

THEOREM 2.1: $SW \cap J(S) = (W \cap J(R)) \oplus KW$.

PROOF: Obviously, $(W \cap J(R)) \oplus KW \subseteq SW$ and $KW \subseteq J(S)$. Let $r \in W \cap J(R)$ and $s = \tau + k \in S$, for some $\tau \in R$ and $k \in K$. Put $\beta = sr = \tau r + kr$. Since $\tau r \in J(R)$ there exists a left quasi-inverse $b \in R$ such that $b*(\tau r) = \tau r + b + b\tau r = 0$ and consequently $b*\beta = kr + bkr = \gamma \in J(S)$. That, in turn, yields the existence of $d \in S$ such that $d*\gamma = 0$ and since the "star" composition $*$ is associative, $(d*b)*\beta = 0$. Therefore, $r \in J(S)$ and we obtain $(W \cap J(R)) \oplus KW \subseteq SW \cap J(S)$.

Conversely, since $W \cap J(S) \subseteq R \cap J(S) \subseteq J(R)$, we obtain $SW \cap J(S) = (W \oplus KW) \cap J(S) = (W \cap J(S)) \oplus KW \subseteq (W \cap J(R)) \oplus KW$. □

EXAMPLE 2.2: If $S = R[x;\alpha]$, where α is an automorphism of R, then $W = \{r \in R;\ xr \in J(S)\}$ and $K = \{\Sigma\ x^i r_i,\ r_i \in R \text{ and } i \geq 1\}$. Furthermore, $J(S) = (W \cap J(R)) \oplus KW$,[2].

THEOREM 2.3: Suppose that $J(R)S \subseteq SJ(R)$. Then either of the following conditions implies $J(R) \subseteq J(S)$:

(i) $J(R)$ is right T-nilpotent;

(ii) S_R is finitely generated;

(iii) simple right S-modules are R-projective.

PROOF: Let M_S be a simple S-module. Thanks to $J(R)K \subseteq SJ(R)$, $MJ(R)$ is an S-submodule of M_S. Since either of the three conditions (i), (ii) or (iii) implies that $MJ(R) \neq M$ ([1], pp. 198, 314]) we obtain $MJ(R) = 0$. Hence $J(R) \subseteq J(S)$. □

(Notice that the theorem holds for arbitrary extensions with the same identity.)

THEOREM 2.4: Suppose $M + K$ is a right S-ideal for each maximal right ideal $M \subseteq R$ (e.g. $MK \subseteq M + K$ and $K^2 \subseteq J(R) + K$). Then $J(S) \subseteq J(R) \oplus K$. Moreover, if K is nil modulo J(R) then $J(S) = J(R) \oplus K$.

PROOF: If $M \oplus K$ is a right S-ideal then $J(S) \subseteq \cap (M \oplus K) = J(R) \oplus K$, where the intersection runs through all maximal $M_R \subseteq R_R$. Now, assume that K is nil modulo J(R), i.e. for each $k \in K$ there exists a natural number n such that $k^n \in J(R)$. Let $j + k \in J(R) \oplus K$ and $s \in S$. According to the hypothesis $(j+k)s = j' + k'$, where $j' \in J(R)$, i.e. $(1 - (j+k)s) = (1 - j') - k$.

Since $j' \in J(R)$ there exists $r \in R$ such that $(1 - j')r = 1$, i.e. $(1 - (j+k)s)r = 1 - k'r$, and $k'r \in K$. Now, $(k'r)^n \in J(R)$, for some n, and hence $(1 - (k'r)^n)r' = 1$, for some $r' \in R$. However,

$$1 = (1-(k'r)^n)r' = (1-k'r)\left(\sum_{i=0}^{n-1} (k'r)^i\right)r'$$

yields that $(1-k'r)$ has a right inverse. Thus $(j+k) \in J(S)$. □

The following two theorems provide a generalization and an improvement of the normalizing basis theorem ([9], p. 276).

THEOREM 2.5: Suppose K_R is projective. Then:

(i) If $\mathrm{Hom}_R({}_RS_R, M_R)_R$ is of finite length n_M for every simple M_R then $J^w(S) \subseteq SJ(R)$, where $w = \sup\{n_M\}$.

(ii) If K_R is finitely generated free then either of the following conditions implies that $\mathrm{Hom}_R({}_RS_R, M_R)_R$ is of finite length, for every simple M_R:

(a) each maximal $N_R \subseteq R_R$ is an ideal of R and NS = SN;

(b) R/J(R) is artinian and SJ(R) = J(R)S;

(c) $S_R = \sum_{i=1}^{n} x_i R$ is a free module with basis $\{x_i;\ i = 1,\dots,n\}$; $x_1 = 1$;

and for every $r \in R$, there are $r_i \in R$, $1 \leq i < n$, $r_1 = r$, such that

$$rx_j = (\sum_{i=1}^{j-1} x_i r_i) + x_j \sigma_j(r), \quad 1 < j \leq n,$$

where σ_j's are ring automorphisms of R.

(TRIANGULAR MATRIX COMMUTATION)

PROOF: (i) If $U_R = \mathrm{Hom}_R({}_RS_R, M_R)_R$ is of finite length then $U_S = \mathrm{Hom}_R({}_SS_R, M_R)_S$ is of finite length, too, and length$(U_S) \leq n_M$. Therefore, $U(J^k(S)) = 0$, where $k = n_M$. Since K_R is projective, S_R is projective, too, and $J_R(S_R) = S(J(R))$ ([1], p. 196). On the other hand, $U(J^k(S)) = 0$ yields that $\mathrm{Hom}_R({}_SS_R, M_R)_S(J^w(S)) = 0$, for every simple M_R, and that in turn implies $J^w(S) \subseteq J_R(S_R)$. (Define $J^w(S) = \cap J^n(S)$, where the intersection runs through $n \in \{n_M\}$.)

(ii) (a) Let $M_R \simeq R/N$, where $N_R \subseteq R_R$ is maximal. Then $M \otimes {}_RS_R \simeq (S/NS)_R = (S/SN)_R$ is a homogeneous semisimple R-module of finite length. Since $(\mathrm{Hom}_R({}_RS_R, M_R)_R)N = 0$, $\mathrm{Hom}_R({}_RS_R, M_R)_R$ is semisimple and homogeneous. Now, the natural homomorphism ${}_E\mathrm{Hom}_R(M \otimes {}_RS_R, {}_EM_R) \simeq {}_E\mathrm{Hom}_R(M_R, {}_E\mathrm{Hom}_R({}_RS_R, {}_EM_R)_R)$, where $E = \mathrm{End}_R(M_R)$ is a skew-field, implies that $\mathrm{Hom}_R({}_RS_R, M_T)_R$ is of finite length, too.

(ii) (b) Again, let $M_R \simeq R/N$, where $N_R \subseteq R_R$ is maximal. Then $M \otimes {}_RS_R \simeq (S/NS)_R$ is isomorphic to a factor module of $(S/(J(R)S))_R = (S/(SJ(R)))_R$ that is semisimple of finite length. Similarly, since $(\mathrm{Hom}_R({}_RS_R, M_R)_R J(R) = 0$ and R/J(R) is artinian semisimple we obtain that $\mathrm{Hom}_R({}_RS_R, M_R)_R$ is semisimple, too. Now, using the natural transformation introduced in the proof above, again we obtain that $\mathrm{Hom}_R({}_FM'_R, \mathrm{Hom}_R({}_RS_R, M_K)_R)_F$ is finite-dimensional over $F = \mathrm{End}_R(M'_R)$, for every simple M'_R. Since the representative set of simple right R-modules is finite (thanks to R/J(R) being artinian), $\mathrm{Hom}_R({}_RS_R, M_R)_R$ is necessarily of finite length.

(ii) (c) Let $\sum_{i=1}^{n} x_i u_i \in S_R$. Then $r(\sum_{i=1}^{n} x_i u_i) = \sum_{j=1}^{n} x_j t_j$ and the relationship between u_i's and t_j's can be expressed in an upper triangular matrix form by

$$\begin{vmatrix} t_1 \\ \\ \vdots \\ t_n \end{vmatrix} = \begin{vmatrix} r & x & x & \cdots & x & x \\ 0 & \sigma_2(r) & x & \cdots & x & x \\ \vdots & \vdots & & & & \vdots \\ 0 & 0 & & \cdots & 0 & \sigma_n(r) \end{vmatrix} \begin{vmatrix} u_1 \\ \\ \vdots \\ u_n \end{vmatrix} .$$

Consequently, there is a ring monomorphism $\Psi:R \to UT_n(R)$, where $UT_n(R)$ is the ring of upper triangular matrices of the size n with entries from R, such that the diagonal entries of $\Psi(r)$ are given by $(\Psi(r))_{ii} = \sigma_i(r)$, $i = 1,\ldots,n$, (define σ_1 = identity). Now, let $f \in \mathrm{Hom}_R({}_RS_R, M_R)_R$ and $f(x_i) = m_i$, $i = 1,\ldots,n$. Then

$$(fr)(\sum_{i=1}^{n} x_i r_i) = f(\sum_{j=1}^{n} m_j t_j)$$

and we can view $\mathrm{Hom}_R({}_RS_R, M_R)_R \simeq (M_R^n)_R$ as a direct sum $(M_1 \oplus \ldots \oplus M_n)_R$, where $M = M_i$, $i = 1,\ldots,n$, with the scalar right R-multiplication being accomplished by right matrix multiplication with elements of $\Psi(R)$. In particular, if $1 \leqq k \leqq n$, then for every $m_k \in M_k$, $(0,\ldots,m_k,0,\ldots,0)r = (0,\ldots,0,m_k\sigma_k(r),m'_{k+1},\ldots,m'_n)$, for some $m'_i \in M_i$, $i = k+1,\ldots,n$. Hence $(M_k \oplus \ldots \oplus M_n)$ is an R-submodule of $(M_1 \oplus \ldots \oplus M_n)_R$ for every $1 \leqq k \leqq n$, and since M_R is simple, $(M_k \oplus \ldots \oplus M_n)R = (0,\ldots,0,m_k,0,\ldots,0)R + (M_{k+1} \oplus \ldots \oplus M_n)R$, for each $0 \neq m_k \in M$. Furthermore, σ_k being a ring automorphism implies that $\{r \in R;\ m_k r \in (M_{k+1} \oplus \ldots \oplus M_n)\} = \sigma_k^{-1}\{r \in R\ ;\ m_k r = 0\}$, a maximal right ideal of R. Thus $(M_k \oplus \ldots \oplus M_n)/(M_{k+1} \oplus \ldots \oplus M_n))_R$ is simple for each $1 \leqq k \leqq n$, and consequently $(M_1 \oplus \ldots \oplus M_n)_R$ is of finite length n. □

<u>THEOREM 2.6</u>: If $\mathrm{Hom}_R({}_RS_R, M_R)_R$ is semisimple for each simple M_R and every S-submodule that is an R-direct summand of an S-module is also an S-direct summand, then $J(S) \subseteq J(R) \oplus J_R(K_R)$. Either of the following conditions

implies that $\mathrm{Hom}_R({}_RS_R, M_R)_R$ is semisimple, for every simple M_R:

(i) each maximal $N_R \subseteq R_R$ is an ideal of R and $NS \subseteq SN$;

(ii) $R/J(R)$ is artinian and $J(R)S \subseteq SJ(R)$;

(iii) $S_R = \Sigma\, x_iR$ is a free module with basis $\{x_i;\ i \in \Lambda\}$ and for every $r \in R$, $rx_i = x_i\sigma_i(r)$, $i \in \Lambda$, where σ_i's are ring automorphisms of R; and either K_R is finitely generated or $R/J(R)$ is artinian.

(DIAGONAL MATRIX COMMUTATION)

PROOF: Let M_R be simple. The hypothesis implies that $\mathrm{Hom}_R({}_SS_R, M_R)_S$ is semisimple, too, and $0 = \mathrm{Hom}_R(S_R, M_R)J(S)$. Therefore, $J(S)S = J(S) \subseteq J_R(S_R) = J(R) \oplus J_R(K_R)$.

For the proofs of (i) and (ii) we can use the same methods as we did in proving Theorem 2.5 (ii)(a) and (b). (Notice that we do not require that $\mathrm{Hom}_R({}_RS_R, M_K)_R$ be of finite length here.)

(iii) If K_R is finitely generated then similarly as in the proof of Theorem 2.5 (ii)(c), R is acting on $\mathrm{Hom}_R({}_RS_R, M_R)_R$ as diagonal matrices

$$\begin{vmatrix} r & 0 & 0 & \dots & 0 & 0 \\ 0 & \sigma_2(r) & 0 & \dots & 0 & 0 \\ \vdots & \vdots & & & & \vdots \\ 0 & 0 & & \dots & 0 & \sigma_n(r) \end{vmatrix},$$

(we set $\sigma_1(r) = r$, for convenience) $\mathrm{Hom}_R({}_RS_R, M_R)_R \simeq (M_1 \oplus \dots \oplus M_n)_R$, where $(M_k)_R \simeq M_R$, for each $k = 1,\dots,n$, and $(0,\dots,0, M_k, 0,\dots,0)_R \simeq M_R$, i.e. $\mathrm{Hom}_R({}_RS_R, M_R)_R$ is semisimple. In general, the "diagonal" commutation hypothesis implies that $J(R)S = SJ(R)$ (since $J(R)$ is stable under ring automorphisms of R). Therefore $\mathrm{Hom}_R({}_RS_R, M_R)_RJ(R) = 0$ and $R/J(R)$ being artinian implies that $\mathrm{Hom}_R({}_RS_R, M_K)_R$ is semisimple. □

REFERENCES

[1] F.W. Anderson and K.R. Fuller, Rings and Categories of Modules, Springer, Berlin, 1974.

[2] S.S. Bedi and J. Ram, Jacobson radical of skew polynomial rings and skew group rings, Israel J. Math. 35 (1980), 327-338.

[3] S. Bose, Splitting of ring extensions, Indian J. Pure Appl. Math. 16 (1983), 355-356.

[4] P.M. Cohn, Free Rings and Their Relations, Academic Press, New York, 1971.

[5] C. Faith, Algebra, Rings, Modules and Categories I, Springer, Berlin, 1973.

[6] K.R. Goodearl, Ring Theory: Nonsingular Rings and Modules, Marcel Dekker, New York, 1976.

[7] N. Jacobson, Basic Algebra II, W.H. Freeman, San Francisco, 1980.

[8] I. Palmer, The global homological dimension of semi-trivial extensions of rings, Math. Scand. 37 (1975), 223-256.

[9] D.S. Passman, The Algebraic Structure of Group Rings, Wiley, New York, 1977.

P.E. Jambor
Department of Mathematics,
University of North Carolina,
Wilmington,
NC 28406
U.S.A.

SHOJI KYUNO

A nonunital Morita ring and its radicals

ABSTRACT: The theorem of Morita on correspondences of modules over unital rings was generalized in the case of nonunital rings in [3], [4] and [6]. Thus, a nonunital Morita ring became the object of our interest. At first, various nonunital Morita rings are given and we show that nonunital Morita rings are essentially identical with weak Nobusawa gamma rings. The concept of closed ideals was introduced in [5] and studied in [2]. Using the concept of upper closed ideals, we characterize the simplicial radical of a nonunital Morita ring. The results give generalizations of similar ones obtained in [1].

1. NONUNITAL MORITA RINGS

DEFINITION 1.1: Let R, L be rings and M,Γ additive abelian groups. If

$$C = \begin{pmatrix} R & \Gamma \\ M & L \end{pmatrix} = \{ \begin{pmatrix} r & \gamma \\ m & \ell \end{pmatrix} \mid r \in R, \gamma \in \Gamma, m \in M, \ell \in L\}$$

forms a ring under the usual definitions of matrix addition and multiplication, C is called a *Morita ring*.

From the definition we see immediately that M is an L-R-bimodule and Γ is an R-L-bimodule. In this note, we assume $\Gamma M = R$ and $M\Gamma = L$. It is easy to see that C has a unity if and only if R and L have unities and both M and Γ are unitary. If C does not have a unity, the Morita ring C is called a *nonunital Morita ring*. The following are examples of nonunital Morita rings:

EXAMPLES:

1. $\begin{pmatrix} ReR & Re \\ eR & eRe \end{pmatrix}$

where R is a ring such that $R^2 = R$ and $e^2 = e$.

2. $\begin{pmatrix} R & R_{1\infty} \\ R'_{\infty 1} & R'_{\infty\infty} \end{pmatrix}$

where R is a ring such that $R^2 = R$, $R_{1\infty}$ denotes the set of all row vectors over R and $R'_{\infty 1}$ denotes the set of all column vectors with at most finite entries except 0 over R and $R'_{\infty\infty} = R'_{\infty 1}R_{1\infty}$.

3. $\begin{pmatrix} BA & B \\ A & AB \end{pmatrix}$

where R is a ring and A, B are ideals of R.

4. $\begin{pmatrix} M^*M & M^* \\ M & MM^* \end{pmatrix}$

where M_R is a right R-module and $M^* = \mathrm{Hom}_R(M,R)$.

5. $\begin{pmatrix} NM & N \\ M & MN \end{pmatrix}$

where A, B are additive abelian groups, $M \subseteq \mathrm{Hom}(A,B)$ and $N \subseteq \mathrm{Hom}(B,A)$ are subgroups such that $MNM \subseteq M$ and $NMN \subseteq N$.

6. $\begin{pmatrix} \Gamma M & \Gamma \\ M & M\Gamma \end{pmatrix}$

where (Γ,M) is a weak Nobusawa gamma ring, ΓM denotes $\Gamma \otimes_Z M$ and $M\Gamma$ denotes $M \otimes_Z \Gamma$.

<u>REMARK 1.2</u>: Let M and Γ be additive abelian groups. If for all $a,b,c \in M$ and $\alpha,\beta,\gamma \in \Gamma$ the conditions:

(N_1) $a\alpha b \in M$, $\alpha a\beta \in \Gamma$;

(N_2) $(a+b)\alpha c = a\alpha c+b\alpha c$, $(\alpha+\beta)a\gamma = \alpha a\gamma+\beta a\gamma$, $a(\alpha+\beta)b = a\alpha b+a\beta b$,
$\alpha(a+b)\beta = \alpha a\beta+\alpha b\beta$, $a\alpha(b+c) = a\alpha b+a\alpha c$, $\alpha a(\beta+\gamma) = \alpha a\beta+\alpha a\gamma$;

(N_3) $(a\alpha b)\beta c = a(\alpha b\beta)c = a\alpha(b\beta c)$,
$(\alpha a\beta)b\gamma = \alpha(a\beta b)\gamma = \alpha a(\beta b\gamma)$

are satisfied, then (Γ,M) is called a *weak Nobusawa gamma ring*. We denote an element of ΓM by $\Sigma_i\gamma_i x_i$. Similarly, $\Sigma_j y_j\beta_j$ denotes an element of $M\Gamma$. Define $\Sigma_i\gamma_i x_i\Sigma_j\beta_j y_j = \Sigma_{i,j}\gamma_i(x_i\beta_j y_j)$, where $\Sigma_i\gamma_i x_i$, $\Sigma_j\beta_j y_j \in \Gamma M$, and $x\Sigma_i\gamma_i x_i = \Sigma_i x\gamma_i x_i$, $x \in M$ and $\gamma\Sigma_j y_j\beta_j = \Sigma_j\gamma y_j\beta_j$, $\gamma \in\Gamma$. If A is an L-R-submodule of M, then A is called an *ideal* of M and denoted by $A \triangleleft M$. Similarly for $D \triangleleft \Gamma$, where D is an R-L-submodule of Γ.

Some more examples of nonunital Morita rings are:

7. $\begin{pmatrix} \Gamma BM & \Gamma \\ BM & BM\Gamma \end{pmatrix}$

where (Γ,M) is a weak Nobusawa gamma ring and B is an ideal of $M\Gamma$.

8. $\begin{pmatrix} R_{n,n} & \Gamma_{n,m} \\ M_{m,n} & L_{m,m} \end{pmatrix}$

where (Γ,M) is a weak Nobusawa gamma ring, $R = \Gamma M$, $L = M\Gamma$ and $M_{m,n}$ is the set of all $m \times n$ matrices over M, etc.

From Example 6 it is clear that for any weak Nobusawa gamma ring (Γ,M) we can obtain a Morita ring. Conversely, given a Morita ring $\begin{pmatrix} R & \Gamma \\ M & L \end{pmatrix}$, we then obtain a weak Nobusawa gamma ring (Γ,M). It follows that the concept of a Morita ring is essentially identical with that of a weak Nobusawa gamma ring.

2. IDEALS OF A NONUNITAL MORITA RING

Let $C = \begin{pmatrix} R & \Gamma \\ M & L \end{pmatrix}$, $R = \Gamma M$, $L = M\Gamma$, be a nonunital Morita ring. Let A be an ideal of L. If we define

$$A\Gamma^{-1} = \{m \in M \mid m\Gamma \subseteq A\},\quad M^{-1}A = \{\gamma \in \Gamma \mid M\gamma \subseteq A\},$$

$$M^{-1}A\Gamma^{-1} = \{r \in R \mid Mr\Gamma \subseteq A\},\quad AL^{-1} = \{\ell \in L \mid \ell L \subseteq A\},$$

$$L^{-1}A = \{\ell \in L \mid L\ell \subseteq A\},\quad L^{-1}AL^{-1} = \{\ell \in L \mid L\ell L \subseteq A\},$$

then $A\Gamma^{-1} \triangleleft M$, $M^{-1}A \triangleleft \Gamma$, $M^{-1}A\Gamma^{-1} \triangleleft R$, $AL^{-1} \triangleleft L$, $L^{-1}A \triangleleft L$ and $L^{-1}AL^{-1} \triangleleft L$.

Furthermore,

$$\begin{pmatrix} M^{-1}A\Gamma^{-1} & M^{-1}A \\ A\Gamma^{-1} & A \end{pmatrix} \lhd \begin{pmatrix} R & \Gamma \\ M & L \end{pmatrix}.$$

DEFINITION 2.1: A mapping Φ from the set of ideals in L to the set of ideals in L is defined by the following:

$$\Phi : I(L) \to I(M) \longrightarrow I(R) \longrightarrow I(\Gamma) \longrightarrow I(L)$$

$$A \longmapsto A\Gamma^{-1} \longmapsto M^{-1}A\Gamma^{-1} \longmapsto M^{-1}A\Gamma^{-1}M^{-1} \longmapsto \Gamma^{-1}M^{-1}A\Gamma^{-1}M^{-1},$$

where $A \lhd L$ and $I(X)$ denotes the set of ideals in X.

DEFINITION 2.2: Let L be a ring and A be an ideal of L. If $L^{-1}AL^{-1} = A$, then A is called an *upper closed ideal*.

It is easy to see that prime ideals are upper closed. Furthermore, if $L^2 = L$ then maximal ideals in L are upper closed, and if $A_i \lhd L$ $(i \in I)$ are upper closed then $\bigcap_{i \in I} A_i$ is upper closed.

PROPOSITION 2.3 [2, Theorem 3]: Let I be an ideal of $C = \begin{pmatrix} R & \Gamma \\ M & L \end{pmatrix}$. Then there exists an ideal A of C such that

$$A = \begin{pmatrix} A_{11} & A_{12} \\ A_{21} & A_{22} \end{pmatrix}$$

and $I \subseteq A \subseteq C^{-1}I\,C^{-1}$, where $A_{11} \lhd R$, $A_{12} \lhd \Gamma$, $A_{21} \lhd M$ and $A_{22} \lhd L$.

PROPOSITION 2.4 [2, Theorem 4]: Let I be an ideal of $C = \begin{pmatrix} R & \Gamma \\ M & L \end{pmatrix}$. Then I is upper closed if and only if

$$I = \begin{pmatrix} I_{11} & I_{12} \\ I_{21} & I_{22} \end{pmatrix}$$

where each I_{ij} is upper closed and the relations between the I_{ij}'s are given by Φ.

3. THE SIMPLICIAL RADICAL OF A NONUNITAL MORITA RING

Let $S(C)$ denote the simplicial radical of C, that is, the intersection of all maximal ideals in C.

We first prove the following lemma.

LEMMA 3.1: Let $C^2 = C$. An ideal I of C is a maximal ideal if and only if

$$I = \begin{pmatrix} M^{-1}B\Gamma^{-1}\Gamma & M^{-1}B \\ B\Gamma^{-1} & B \end{pmatrix},$$

where B is a maximal ideal of L.

PROOF: Let B be a maximal ideal of L. Put

$$I = \begin{pmatrix} M^{-1}B\Gamma^{-1} & M^{-1}B \\ B\Gamma^{-1} & B \end{pmatrix}.$$

We show that I is a maximal ideal of C.

Let J be an ideal of C such that $I \subseteq J \subseteq C$. By Proposition 2.3, there exists an ideal $\begin{pmatrix} J_{11} & J_{12} \\ J_{21} & J_{22} \end{pmatrix}$ such that

$$J \subseteq \begin{pmatrix} J_{11} & J_{12} \\ J_{21} & J_{22} \end{pmatrix} \subseteq C^{-1}J\ C^{-1}.$$

Since B is maximal, $B = J_{22}$ or $J_{22} = L$. If $B = J_{22}$, $MJ_{12} \subseteq B$ implies $J_{12} \subseteq M^{-1}B$, $J_{21} \subseteq B$ implies $J_{21} \subseteq B\Gamma^{-1}$, and $MJ_{11}\Gamma \subseteq B$ implies $J_{11} \subseteq M^{-1}B\Gamma^{-1}$. Thus,

$$J \subseteq \begin{pmatrix} J_{11} & J_{12} \\ J_{21} & J_{11} \end{pmatrix} \subseteq I$$

and so $J = I$. If $J_{22} = L$, $J_{22}M \subseteq J_{21}$ implies $M = M\Gamma M = LM \subseteq J_{21}$, and so $J_{21} = M$. Here, $M = M\Gamma M$ follows from $C^2 = C$. Then, $\Gamma M = \Gamma J_{21} \subseteq J_{11}$ implies $J_{11} = R$. Since $\Gamma = \Gamma M \Gamma$ follows from $C^2 = C$, $\Gamma = \Gamma M\Gamma = \Gamma L = \Gamma J_{22} \subseteq J_{12}$ implies $J_{12} = \Gamma$. Thus, $\begin{pmatrix} J_{11} & J_{12} \\ J_{21} & J_{22} \end{pmatrix} = C$. Since $C \begin{pmatrix} J_{11} & J_{12} \\ J_{21} & J_{22} \end{pmatrix} C \subseteq J$, $C = C^3 \subseteq J$ implies $J = C$. Therefore, C is maximal.

Conversely, let I be a maximal ideal of C. Since I is upper closed, it follows from Proposition 2.4 that

$$I = \begin{pmatrix} M^{-1}B\Gamma^{-1} & M^{-1}B \\ B\Gamma^{-1} & B \end{pmatrix}$$

where B is an upper closed ideal of L. We show that B is maximal. Assume that B is not maximal. Then there exists an ideal A in L such that $B \subsetneq A \subsetneq L$. Then the ideal

$$J = \begin{pmatrix} M^{-1}A\Gamma^{-1} & M^{-1}A \\ A\Gamma^{-1} & A \end{pmatrix}$$

satisfies $I \subsetneq J \subsetneq C$, a contradiction. □

It is easy to verify that Φ preserves the maximality of ideals between L, M, R and Γ. That is, if B is a maximal ideal in L then $B\Gamma^{-1}$ is a maximal ideal in M; if A is a maximal ideal in M then $M^{-1}A$ is a maximal ideal in R; and so on. Thus, we have

$$\Phi : B \mapsto B\Gamma^{-1} \mapsto M^{-1}B\Gamma^{-1} \mapsto M^{-1}B\Gamma^{-1}M^{-1} \mapsto \Gamma^{-1}M^{-1}B\Gamma^{-1}M^{-1} = L^{-1}B\,L^{-1}$$

where all ideals in the sequence are maximal.

If B is a maximal ideal in L, then $BL^{-1} = B$ and $L^{-1}B = B$. Hence, $M^{-1}B\Gamma^{-1}M^{-1} = M^{-1}B$ and $L^{-1}BL^{-1} = B$. Therefore, Φ derives one-to-one correspondences between maximal ideals of L, M, R and Γ.

<u>THEOREM 3.2</u>: If $C^2 = C$, then

$$S(C) = \begin{pmatrix} M^{-1}\,S(L)^{-1} & M^{-1}\,S(L) \\ S(L)\Gamma^{-1} & S(L) \end{pmatrix},$$

where

$$S(R) = M^{-1}\, S(L)\Gamma^{-1}, \quad S(L) = \Gamma^{-1}\, S(R)M^{-1},$$

$$S(\Gamma) = M^{-1}\, S(L) \text{ and } \quad S(M) = S(L)\Gamma^{-1}.$$

PROOF:

$$\begin{aligned}
S(C) &= \bigcap_{I \text{ maximal in } C} I \\
&= \bigcap_{B \text{ maximal in } L} \begin{pmatrix} M^{-1}B\Gamma^{-1} & M^{-1}B \\ B\Gamma^{-1} & B \end{pmatrix} \quad \text{(by Lemma 3.1)} \\
&= \begin{pmatrix} \cap\, M^{-1}B\Gamma^{-1} & \cap\, M^{-1}B \\ \cap\, B\Gamma^{-1} & \cap\, B \end{pmatrix} \\
&= \begin{pmatrix} M^{-1}\, S(L)\Gamma^{-1} & M^{-1}\, S(L) \\ S(L)\Gamma^{-1} & S(L) \end{pmatrix} .
\end{aligned}$$

So $S(R) = M^{-1}\, S(L)\Gamma^{-1}$, $S(L) = \Gamma^{-1}\, S(R)M^{-1}$, $S(\Gamma) = M^{-1}\, S(L)$ and $S(M) = S(L)\Gamma^{-1}$. Each equality follows from the one-to-one correspondence between maximal ideals determined by the mapping Φ. □

COROLLARY 3.3: If $C^2 = C$, then

$$C/S(C) \cong \begin{pmatrix} R/S(R) & \Gamma/S(\Gamma) \\ M/S(M) & L/S(L) \end{pmatrix}.$$

PROOF: The mapping

$$\begin{pmatrix} r & \gamma \\ m & \ell \end{pmatrix} + S(C) \longmapsto \begin{pmatrix} r + S(R) & \gamma + S(\Gamma) \\ m + S(M) & \ell + S(L) \end{pmatrix}$$

is a ring isomorphism. □

REMARK 3.4: By applying a similar procedure as in the case of $S(C)$, we get the following for the prime radical $P(C)$ without assuming $C^2 = C$:

$$P(C) = \begin{pmatrix} M^{-1}P(L)\Gamma^{-1} & M^{-1}P(L) \\ P(L)\Gamma^{-1} & P(L) \end{pmatrix},$$

where $P(R) = M^{-1}P(L)\Gamma^{-1}$, $P(L) = \Gamma^{-1}P(R)M^{-1}$, $P(\Gamma) = M^{-1}P(L)$ and $P(M) = P(L)\Gamma^{-1}$.

For the Jacobson radical $J(C)$, we define a C-module and an irreducible C-module as follows. If for a pair C and $\begin{pmatrix} W_1 \\ W_2 \end{pmatrix}$, where W_1, W_2 are additive abelian groups, the product

$$C\begin{pmatrix} W_1 \\ W_2 \end{pmatrix} = \begin{pmatrix} RW_1 + \Gamma W_2 \\ MW_1 + LW_2 \end{pmatrix} \subseteq \begin{pmatrix} W_1 \\ W_2 \end{pmatrix}$$

is defined and $\begin{pmatrix} W_1 \\ W_2 \end{pmatrix}$ satisfies the usual module axioms, $\begin{pmatrix} W_1 \\ W_2 \end{pmatrix}$ is called a *left C-module*. A left C-module is said to be *irreducible* if W_1 isan irreducibl left R-module and W_2 is an irreducible left L-module.

We can verify:

LEMMA 3.5: Let $\begin{pmatrix} W_1 \\ W_2 \end{pmatrix}$ be an irreducible left C-module. Then,

$$\left(0 : \begin{pmatrix} W_1 \\ W_2 \end{pmatrix}\right) = \begin{pmatrix} M^{-1}P\Gamma^{-1} & M^{-1}P \\ P\Gamma^{-1} & P \end{pmatrix},$$

where

$$\left(0 : \begin{pmatrix} W_1 \\ W_2 \end{pmatrix}\right) = \left\{\begin{pmatrix} r & \gamma \\ m & \ell \end{pmatrix} \in C \;\middle|\; \begin{pmatrix} r & \gamma \\ m & \ell \end{pmatrix}\begin{pmatrix} W_1 \\ W_2 \end{pmatrix} = \begin{pmatrix} 0 \\ 0 \end{pmatrix}\right\}$$

and

$$P = (0 : W_2) = \{\ell \in L \mid \ell W_2 = 0\}.$$

LEMMA 3.6: For an irreducible left L-module W_2, there exists an irreducible left R-module W_1 such that $\begin{pmatrix} W_1 \\ W_2 \end{pmatrix}$ is an irreducible left C-module.

Using similar argument as for $S(C)$, we obtain

$$J(C) = \begin{pmatrix} M^{-1}J(L)\Gamma^{-1} & M^{-1}J(L) \\ J(L)\Gamma^{-1} & J(L) \end{pmatrix},$$

where $J(R) = M^{-1}J(L)\Gamma^{-1}$, $J(L) = \Gamma^{-1}J(R)M^{-1}$, $J(\Gamma) = M^{-1}J(L)$ and $J(M) = J(L)\Gamma^{-1}$, without needing C^2 to be equal to C.

These results for $P(C)$ and $J(C)$ follow also from a result on N-radicals by Sands [7].

REFERENCES

[1] S. Kyuno, Nobusawa's gamma rings with the right and left unities, Math. Japonica 25 (1980), 179-190.

[2] S. Kyuno, M.B. Smith and N. Nobusawa, Closed ideals in non-unital matrix rings, Math. J. Okayama Univ. to appear.

[3] S. Kyuno, Equivalence of module categories, Math. J. Okayama Univ. 28 (1986), 147-150.

[4] N. Nobusawa, Γ-rings and Morita equivalences, Math. J. Okayama Univ. 26 (1984), 151-156.

[5] N. Nobusawa, On Morita pairs of rings, Math. J. Okayama Univ. to appear.

[6] N. Nobusawa, Correspondences of modules over a Morita ring, Math. J. Okayama Univ. to appear.

[7] A.D. Sands, Radicals and Morita contexts, J. Algebra 24 (1973), 335-345.

S. Kyuno
Department of Mathematics
Tohoku Gakuin University
Tagajo,
Miyagi 985
Japan

BARBARA L. OSOFSKY

Rings whose idempotent generated ideals form a lattice with ACC

One does not normally expect that the sum of two idempotent generated ideals in a ring R with 1 will be generated by an idempotent. There are two major situations in which this phenomenon occurs. If the idempotents in the ring all commute with each other, one has a Boolean lattice of idempotent generated right ideals. If the ring is von Neumann regular, all finitely generated right ideals are generated by idempotents. In this paper I show that, if a sum of two idempotent generated right ideals is generated by an idempotent, and the ring has acc on idempotent generated right ideals, then the ring is a ring direct product $R_1 \times R_2$, where all idempotents in R_1 are central and R_2 is semisimple artinian and hence von Neumann regular.

Let R be a ring with 1. We will say that R has ISS (for idempotent right ideals sum) if for any $e = e^2$, $f = f^2 \in R$, there is a $g = g^2 \in R$ with $eR + fR = gR$.

We start by looking at some properties of modules related to ISS. These properties are well-known or are comparable to routine exercises in an algebra course, and so will be listed as a series of observations with little in the way of proofs, other than the order in which the observations are given.

Let S be a ring with unit, M_S a unital right S-module, $R = \mathrm{End}_S(M)$. J(S) will denote the Jacobson radical of the ring S.

OBSERVATIONS

(1) There is a one-to-one correspondence between idempotents in R and ordered direct sum decompositions $M = A_S \oplus B_S$ given by

$$e \mapsto eM \oplus (1 - e)M,$$

$$A \oplus B \mapsto \text{the projection of } M \text{ to } A \text{ with kernel } B.$$

(2) Let $M = A \oplus B$, $C \subseteq M$, $M = (A \cap C) \oplus D$. Then

$A + C = A \oplus (D \cap C)$.

(3) If the set of all direct summands of M is closed under + and $\cap$, then R has ISS.

(4) Assume $M = A \oplus C = B \oplus D$. If $A + B$ is projective and e is the projection of M to C with kernel A, then $e|B$ has image $(A + B) \cap C$ which is projective, and kernel $A \cap B$, so $A \cap B$ is a direct summand of $A + B$.

(5) R has ISS $\Longleftrightarrow$ $\{eR \mid e^2 = e \in R\}$ forms a lattice under $\vee = +$ and $\wedge = \cap$ $\Longleftrightarrow$ $\{A \subseteq M \mid M = A \oplus C$ for some $C\}$ forms a lattice under $\vee = +$ and $\wedge = \cap$.

(6) If M is projective, a sum of two direct summands of M is projective $\Longleftrightarrow$ the intersection of two direct summands of M is a direct summand. However, this does not imply that the sum of two direct summands of M is a direct summand, as T = the (hereditary) ring of upper triangular matrices over a field shows. Indeed, let $e,f \in T$, $[e]_{2,2} = [f]_{1,2} = [f]_{2,2} = 1$, all other entries = 0. Then e and f are idempotent, and $eT + fT$ has a direct summand (matrices with zero entries everywhere but the 1,j slot where $j \geq 2$) contained in J(T).

(7) Let M_S be injective. If for all direct summands A and B of M, $A \cap B$ is injective, then for all direct summands A and B of M, $A + B$ is injective. This is the case if and only if for any submodule $K \subseteq M$, K has a unique (as a set) injective hull in M. Equivalently, since idempotents lift modulo J(R) and idempotents with the same kernel mapping onto injective hulls of K are congruent modulo J(R), R has ISS if and only if idempotents lift uniquely modulo J(R), which implies that the lattice of idempotent generated right ideals of R is isomorphic to the lattice of finitely generated right ideals of the von Neumann regular ring R/J(R).

(8) M_S may be injective and have the sum of any two injective submodules injective without R having ISS. Let $S = \Pi_{i \in \omega} F_i$ where the F_i are fields, I a maximal ideal of S containing the direct sum $\oplus_{i \in \omega} F_i$, $K = S/I$, $M = S \oplus K$. Then M is finitely generated injective, and a submodule of M is injective

if and only if it is finitely generated since a finitely generated submodule of S_S is both injective and projective. Thus a sum of two injective submodules of M is injective. However, $(1,0) \cap (1,\bar{1})S$ is not injective. Hence R is not ISS.

By the above observations, if we wish to study the lattice properties of direct summands of a module M_S whose direct summands form a lattice under + and $\cap$, it is sufficient to study the properties of its endomorphism ring, which has ISS. So for the rest of this paper we will look at rings R with ISS. We again start with some observations.

(9) Let $e = e^2$, $f = f^2 \in R$. Then

$$eR = fR \Longleftrightarrow f = e + er(1 - e)$$

for some $r \in R$.

(10) An idempotent $e \in R$ is central

$$\Longleftrightarrow eR(1 - e) = (1 - e)Re = 0$$

$\Longleftrightarrow$ for all right ideals A of R, $R = eR \oplus A$ implies $A = (1 - e)R$ and $R = (1 - e)R \oplus A$ implies $A = eR$.

(11) Let $e = e^2 \in R$. Then for all $r \in R$, $f = e + (1 - e)re$ is idempotent, and $eR + fR = eR \oplus (1 - e)reR$. If $eR + fR$ is generated by an idempotent, then so is $(1 - e)reR$. If I is a two-sided ideal containing no idempotents (such as $J(R)$), then R has ISS $\Rightarrow eI = Ie$ for all $e = e^2 \in R$.

(12) Let R be any ring, $e = e^2$, $f = f^2 \in R$. Then

$$\mathrm{Hom}_R(eR, fR) \approx fRe,$$

and $eR \approx fR \Longleftrightarrow \exists\ \sigma \in eRf$, $\tau \in fRe$, such that $\sigma\tau = e$ and $\tau\sigma = f$.

(13) Let R be a ring with acc on idempotent generated right ideals. Then by

maximum condition, $R = \Pi_{i=1}^{m} R_i$ (ring direct product), where each R_i has no central idempotents other than 0 and 1, and $R_R = \oplus_{i=1}^{n} e_iR$, where each e_i is a primitive idempotent.

We are now ready to classify rings with ISS and acc on idempotent generated right ideals.

<u>LEMMA A</u>: Let R have ISS. Then for all $e = e^2 \in R$, $eR(1 - e) \neq 0 \Longleftrightarrow (1 - e)Re \neq 0$.

<u>PROOF</u>: Assume $(1 - e)Re \neq 0$. Set $f = e + (1 - e)re$. Then $f^2 = f$ and $eR + fR = gR$ for some $g = g^2 \in R$, so

$$R = eR \oplus (1 - e)reR \oplus (1 - g)R.$$

Let ι be the projection of 1 onto $(1 - e)reR$ with respect to this decomposition. Then ι is a nonzero idempotent, and $\iota = (1 - e)res$ for some $s \in R$. Then $es(1 - e) \neq 0$. Replace e by $(1 - e)$ for the reverse implication. □

<u>LEMMA B</u>: Let R be an ISS ring, $e = e^2$, $f = f^2 \in R$, $fe = 0$, $eRf \neq 0$. If e is primitive, i.e. eR is indecomposable, then fR contains a direct summand $\approx eR$.

<u>PROOF</u>: By hypothesis, there is an $r \in R$ such that $erfR \neq 0$. By observation (4), erfR is generated by an idempotent. Since e is primitive, $eR = erfR$. Then the map $fs \mapsto erfs$ maps fR onto eR which is projective. This map must split. □

<u>LEMMA C</u>: Let R have ISS, and let e be a primitive idempotent of R. Then if e is not central, eR is simple.

<u>PROOF</u>: By Lemma A and Observation (10), $eR(1-e) \neq 0$. By Lemma B, $(1 - e)R$ contains a direct summand $fR \approx eR$. By Observation (12), there exist $\sigma \in eRf$ and $\tau \in fRe$ such that $\sigma\tau = e$ and $\tau\sigma = f$. Let $0 \neq r \in eR$. Then $eR \supseteq er(1 - e)R$ which is generated by an idempotent by Observation (11). Since eR is indecomposable, either $rR \supseteq er(1 - e)R = eR$, or $r = re = r\sigma\tau$

and $r\sigma(1 - e)R = eR$. In either case, $rR = eR$, so eR is simple. □

THEOREM: Let R be an associative ring with 1. The following conditions are equivalent:

(a) $R \approx \Pi_{i=1}^{n} R_i$ (ring direct product), where each R_i either has no nontrivial idempotents (other than 0 and 1), or is a simple artinian ring ($M_k(F)$ for some division ring F).

(b) The idempotent generated right ideals of R form a lattice under $\vee = +$ and $\wedge = \cap$, and this lattice has acc.

(c) R is ISS with acc on idempotent generated right ideals.

PROOF: The implications (a) ⇒ (b) ⇒ (c) are straightforward. Now assume (c). By Observation (13) $R \approx \Pi_{i=1}^{n} R_i$, where each R_i contains no central idempotents other than 0 and 1. We thus reduce to the case where $R - \{0,1\}$ is nonempty but contains no central idempotent. By Observation (13), $R_R = \oplus_{i=1}^{n} e_iR$, where each e_i is noncentral and primitive. By Lemma C, each e_iR is simple, so R is (semi)simple artinian. □

There are several consequences of the theorem that lead to interesting questions. We observe that if R has ISS and acc, then a sum of two idempotent generated left ideals is idempotent generated, the lattice of idempotent generated right ideals of R is coordinatized by a von Neumann regular ring $S \times B$ where S is semisimple artinian and B is the boolean algebra of subsets of a finite set (this just means that the lattice of idempotent generated right ideals of R is isomorphic to the lattice of finitely generated right ideals of $S \times B$), and if there are no central idempotents R itself is von Neumann regular. It is interesting to ask whether ISS without any chain condition implies these results. Our final series of observations show that the answer to the left-right symmetry question is yes, and make very small contributions to answering the other two questions.

OBSERVATION (14): Let R have ISS. Then the sum of two idempotent generated left ideals of R is idempotent generated.

PROOF: Let R have ISS, $e = e^2$, $f = f^2 \in R$. One checks that $Re + Rf = Re \oplus Rf(1 - e)$ and $(1 - f)R + (1 - e)R = (1 - f)R \oplus f(1 - e)R$. By ISS, there is an $r \in R$ with $f(1 - e)r$ an idempotent generator of $f(1 - e)R$. Then just as for von Neumann regular rings, $(1 - e)rf(1 - e)$ is an idempotent generator for $Rf(1 - e)$ and $e + (1 - e)rf(1 - e)$ an idempotent generating $Re + Rf$.

OBSERVATION (15): Von Neumann regular rings were introduced to coordinatize many modular, complemented lattices. It seems very likely that they also coordinatize the lattice of idempotent generated right ideals of a ring with ISS, as is the case if the ring is self-injective by Observation (7). To give an idea of the difficulty of using lattice properties only to characterize lattices of finitely generated ideals of von Neumann regular rings, try to distinguish between the lattices given below using only lattice theoretic properties and no counting.

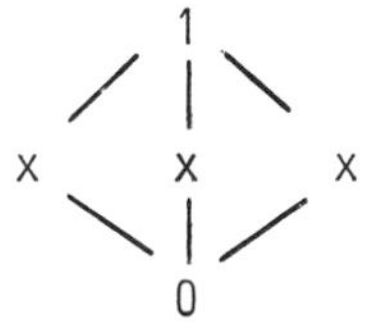

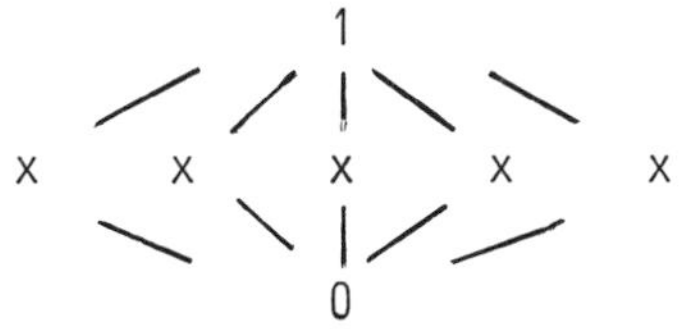

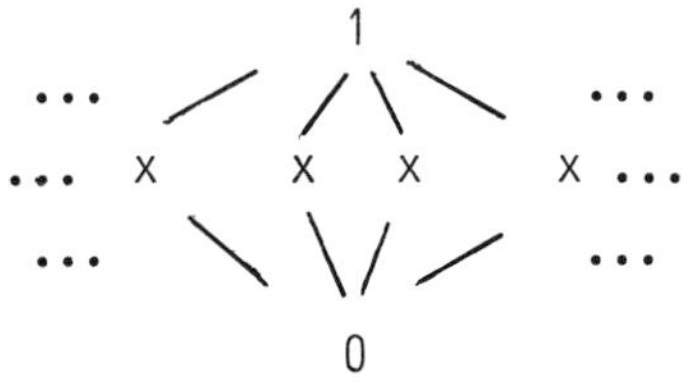

```
              1
            / | \
          X   X   X                    (*)
            \ | /
              0

              1
        /   / | \   \
      X   X   X   X   X                (**)
        \   \ | /   /
              0

                 1
    ...     /   / \   \     ...
    ... X     X     X     X ...        (***)
    ...     \   \ /   /     ...
                 0
```

where there are an infinite number κ of elements in that middle row.

Then (*) is the lattice of idempotent generated right ideals of $M_2(GF(2))$, and (***) the lattice of idempotent generated right ideals of $M_2(F)$ where F is a field of cardinality κ, but (**) cannot be such a lattice by the Theorem.

We conclude with some very minor remarks:

REMARK (16): Let R have ISS, $e \in R - \{0,1\}$, $e = e^2$. Then $R/J(R)$ simple (as a ring) $\Rightarrow$ $J(R) = 0$, where $J(R)$ is the Jacobson radical of R.

PROOF: ReR and $R(1 - e)R$ map onto nonzero ideals of $R/J(R)$, and so map onto $R/J(R)$ under the natural map. By Nakayama's lemma, $ReR = R(1 - e)R = R$. Then $eJ(R) = eJ(R)R(1 - e)R = 0$ and $(1 - e)J(R) = (1 - e)J(R)ReR = 0$ by Observation (11), so $J(R) = 0$. □

REMARK (17): By Observations (1) and (9), if the lattice of idempotent generated right ideals of a ring with ISS is actually finite, then for any noncentral idempotent e, $eR(1 - e)$ is finite. By Observation (5), the theorem has a translation into a theorem about modules whose direct summands form a lattice under $\vee = +$ and $\wedge = \cap$. M_S has only a finite number of direct summands and has the property that its set of direct summands is closed under + and $\cap$ if and only if M_S is a finite sum $M = M_1 \oplus \ldots \oplus M_n$ where $Hom(M_i,M_j) = 0$ for $i \neq j$, M_i is indecomposable for $2 \leq i \leq n$, and M_1 is a finite abelian group with exponent a product of distinct primes (or $\{0\}$).

REMARK (18): It is not true that if R has ISS but no nontrivial central idempotents, then R is von Neumann regular. Let R_1 be the subring of a full ring of linear transformation on an infinite-dimensional vector space over a field F generated by the transformations of finite rank and $F\iota$, where ι is the identity transformation. Let R be obtained from R_1 by adjoining an element x annihilated by itself and all transformations of finite rank. Then R has ISS and centre generated by ι and x over F. However, the lattice of idempotent generated right ideals of R is still coordinatized by the von Neumann regular ring R_1.

B.L. Osofsky
Department of Mathematics
Rutgers - The State University
New Brunswick, NJ 08903
U.S.A.

A.D. SANDS

On dependence and independence among radical properties

1. INTRODUCTION

In earlier papers [7,8,10,12] relations among certain radical properties have been shown to hold. In this paper we show in section 2 that no further independent relations hold among these properties. We do so by deducing that only 23 possible combinations of these nine relations are possible and then by giving examples to show that all 23 cases do indeed occur. Since, classically, all nilpotent rings were assumed to be radical, these examples of radicals are chosen to satisfy this property also. Thus each of these examples of a radical contains the lower Baer radical β, the smallest radical class containing all nilpotent rings. Since every radical must belong to one of these 23 classes, in many cases we are simply pointing out where a known example of a radical lies. We only give detailed proofs when new examples are needed, or when it is necessary to show that a known example does not possess any additional properties.

After considering radicals containing β we turn in section 3 to radicals contained in β. It follows from results of Armendariz [1] that at most 10 of these 23 classes can now be nonempty. We give examples to show that this is indeed the case. Included in this section is a classification of all strict radicals contained in β.

It would be possible to consider any additional property and its effect upon the 23 classes. While β is the lower radical generated by all nilpotent rings, the corresponding upper radical generated by all nilpotent rings is the idempotent radical η consisting of all rings R with $R^2 = R$. In certain problems in radical theory a natural division occurs at η and we consider in section 4 all radicals containing η and in section 5 all radicals contained in η. The first of these problems is settled. It is shown that exactly nine of the 23 classes are nonempty. However in section 5 an open problem is left. It is not known, for radicals contained in η, whether the left hereditary condition implies the right hereditary condition.

We work in the class of associative rings but do not assume that each ring has an identity element. The fundamental definitions and properties of radicals may be found in Divinsky [3] or Wiegandt [14].

We recall the definitions of the radical properties to be considered. If α is a radical class of rings we denote the corresponding semisimple class of rings by S_α. A radical α is said to be hereditary (left hereditary, right hereditary, strongly hereditary) if $R \in \alpha$ and A being an ideal (a left ideal, right ideal, subring) of R implies that $A \in \alpha$. A radical α is said to be left (right) strong if $R \in S_\alpha$ and A being a radical left (right) ideal of R implies that A = 0. A radical α is said to be left stable (right stable, strict) if $R \in S_\alpha$ and A being a left ideal (right ideal, subring) of R implies that $A \in S_\alpha$.

In general we shall work with conditions on the left. The results and examples for conditions on the right may then be obtained by duality.

For any class C of rings the lower and upper radicals generated by C are denoted by $\ell(C)$ and $u(C)$. We recall that the Tangeman-Kreiling construction [13] shows that these hereditary properties of C are inherited by $\ell(C)$. The dual construction [9] shows similarly that properties of C are inherited by the semisimple class of $u(C)$. Earlier proofs of many of these results had been given.

A-radicals were introduced by Gardner [4] as radicals depending only on the additive structure of the rings. Since this concept does not distinguish between "subring" and "ideal", any A-radical is strict and a hereditary A-radical is strongly hereditary.

For any ring R we denote the underlying abelian group by (R,+). R^+ is used to denote the trivial ring defined on (R,+), i.e. the ring where all products are equal to zero. When dealing with A-radicals we use terms such as p-ring, torsion ring, divisible ring, etc., to refer to the corresponding property of the abelian group (R,+).

(1,R) is used to denote the usual Dorroh extension of a ring R to a ring with identity. Z denotes the ring of integers, Q the field of rational numbers, $\mathbb{R}$ the field of real numbers and GF(q) the Galois field with q elements.

2. RELATIONS AMONG RADICAL PROPERTIES

It is clear that strongly hereditary implies left hereditary which, in turn, implies hereditary. Classical examples are known of radicals which separate these properties, e.g. the Brown-McCoy radical is hereditary but not left or right hereditary, the Jacobson radical is left and right hereditary but not strongly hereditary, the lower Baer radical β is strongly hereditary. These examples date from the 1940s. However it was not until 1982 that Beidar [2], at a conference in Eger, Hungary, announced an example separating left hereditary from right hereditary. Since then Puczyłowski [7], without knowing details of Beidar's example, has given another example. Each of these examples depends on the use of a quite complicated domain. We present here a more straightforward example, but using Puczyłowski's ideas concerning infinite matrix rings.

Puczyłowski considers infinite matrix rings over the relevant domain D. Here we use any field F instead of this specific domain. Let V be a vector space of countably infinite dimension over F and let R be the ring $L_f(V,V)$ of linear transformations of finite rank of V. We shall consider V as row vectors over F, each with only finitely many nonzero entries, and writing mappings on the right, R becomes the ring of infinite matrices with entries from F, each matrix involving only finitely many nonzero columns. The radical γ to be considered is the lower radical generated by all left accessible subrings of R. Thus γ is left hereditary. The set A of matrices of R each with only finitely many nonzero entries is a right ideal of R. We show that γ is not right hereditary by showing that A is not in γ. Suppose that A is in γ. Since A is a simple ring it must be a homomorphic image of some left accessible subring of R. It follows from Puczyłowski [7] that A is an image of a left ideal L or R. Let $b \in L$ with $Lb \neq 0$. Then LbR is a nonzero ideal of the simple ring R and so LbR = R. It follows that the right annihilator K of L in L is the unique maximal ideal of L, since LbL = LbRL = RL = L. Since A is a simple ring we must have $L/K \cong A$.

Let VL = W. Then W is a subspace of V. Let $r \in R$ be such that $Vr \subseteq W$. Since r has finite rank there exists a basis $b_{n+1}, b_{n+2}, \ldots$ for the nullspace of r, extending to a basis $b_1, \ldots, b_n, b_{n+1}, \ldots$ of V such that $b_1r, \ldots, b_nr$ is a basis of Vr. Since, for $1 \leqq j \leqq n$, $b_jr \in VL$ there exist $v_{ij} \in V$ and $\ell_{ij} \in L$ such that $b_jr = \sum_i v_{ij}\ell_{ij}$, and each sum is finite. Define s_{ij} by $b_js_{ij} = v_{ij}$ and $b_ks_{ij} = 0$, $k \neq j$, for each j, $1 \leqq j \leqq n$, and each relevant i. Then $s_{ij} \in R$ and $\ell = \sum_j \sum_i s_{ij}\ell_{ij} \in L$. Then $b_k\ell = 0$, $k > n$ and $b_k\ell = \sum_i v_{ik}\ell_{ik} = b_kr$, $1 \leqq k \leqq n$. It follows that $r = \ell \in L$. Thus L consists of all linear transformations of R which map V into W. Then the right annihilator K of L in L is easily seen to consist of all those mappings which map V into W and W to 0. Consider the mapping from L to $L_f(W,W)$ obtained by considering the restriction of ℓ in L to W. It is clear that this is a surjective ring homomorphism with kernel K. Thus $A \cong L/K \cong L_f(W,W)$. However, if W is finite-dimensional, then $L_f(W,W)$ is isomorphic to a ring of finite matrices over F which cannot be isomorphic to A, while if W has infinite dimension then $L_f(W,W) \cong R$. Again we cannot have $A \cong R$. For example, A has countable right uniform dimension, being a direct sum of the minimal right ideals A_i where, for each i, A_i consists of all matrices whose entries are zero except perhaps in row i. Whereas R has uncountable right uniform dimension; if b_k, $k \in K$, is a basis for the vector space over F of all column vectors then, forming the matrix r_k with b_k as first column and other columns zero, one can see that R is the direct sum of the uncountable family of right ideals r_kR of R. This observation on uniform dimension was made to me, verbally, by K. O'Meara, whom I thank.

It is clear that strict implies left stable, which, in turn, implies left strong. Again well-known radicals may be used to give examples separating these properties. For example, the Jacobson radical is left and right strong but not left or right stable, the upper radical generated by all fields is left and right stable but not strict, any A-radical is strict. An example was given in [9] to show that left strong and right strong are independent. Indeed the example actually shows that left stable does not imply right strong. It follows that left stable and right stable are also independent. However these conditions are not completely independent. Puczyłowski [8] has shown that left stable and right strong together imply right stable.

When conditions on a radical α and on $\mathcal{S}_\alpha$ are taken together, relationships do hold true. In [10] it is shown that either left or right hereditary together with either left or right strong imply that a radical is both left and right hereditary and both left and right strong. The corresponding theorem replacing strong by stable was proved by Stewart and Wiegandt [12]. They showed that such radicals are hereditary A-radicals and so that left or right hereditary together with left or right stable imply strongly hereditary and strict.

In order to determine which combinations of these properties can hold it seemed necessary to draw a Venn diagram (figure 1). It turns out that 23 possible combinations exist. In order to show that they actually occur it seems to be necessary to give 23 examples of radicals showing that each region in the diagram is nonempty.

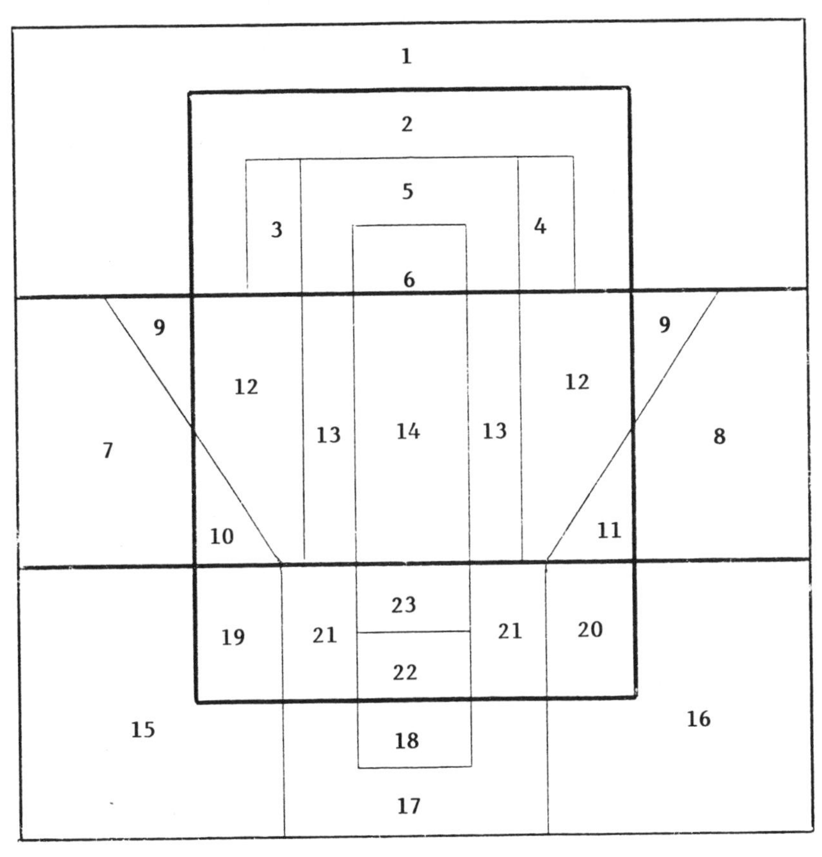

Figure 1

We shall now describe the diagram. In order to preserve left-right symmetry we have found it necessary to split certain regions into two components. The bottom two-thirds of the diagram represents left or right strong, the region between the slanting lines and their continuation having both properties. The bottom third represents left or right stable with the same convention. Puczyłowski's result showing that the intersection of the left stable region and the right strong region is contained in the right stable region and its dual are used here. The small rectangle split into three parts inside the left and right stable region represents strict radicals. The large internal rectangle represents the hereditary region and it intersects all the regions described above. The left and right hereditary regions are contained inside the hereditary region but, from [10], differ only outside the left or right strong regions. Inside it they coincide and are included in the left and right strong region. The top three parts of the narrow central rectangle represent the strongly hereditary radicals. Inside the left or right stable radicals it follows, from [12], that left, right and strongly hereditary coincide and are contained in the strict region.

Thus using the numbering shown in the figure the various properties are represented by the following regions:

Hereditary	$\equiv \{2,3,4,5,6,10,11,12,13,14,19,20,21,22,23\}$;
Left hereditary	$\equiv \{3,5,6,13,14,23\}$;
Right hereditary	$\equiv \{4,5,6,13,14,23\}$;
Strongly hereditary	$\equiv \{6,14,23\}$;
Left strong	$\equiv \{7,9,10,12,13,14,15,17,18,19,21,22,23\}$;
Right strong	$\equiv \{8,9,11,12,13,14,16,17,18,20,21,22,23\}$;
Left stable	$\equiv \{15,17,18,19,21,22,23\}$;
Right stable	$\equiv \{16,17,18,20,21,22,23\}$;
Strict	$\equiv \{18,22,23\}$.

We shall describe a region by its strongest relevant properties. For example, region 19 will be described as hereditary and left stable. Left stable, trivially, implies left strong and we are adopting the convention that no other of these nine properties hold true for radicals in this region. All the examples now given are of radicals for which every nilpotent ring is radical, i.e. radicals containing β.

<u>REGION 1. No properties</u>: Puczyłowski [7] has shown that $\ell(Z,Z^+)$ belongs here.

<u>REGION 2. Hereditary</u>: The Brown-McCoy radical belongs here.

REGION 3. Left hereditary: The first example of such a radical was given by Beidar [2]. The example given earlier here does not contain β, but the lower radical generated by it and β belongs here, as the right ideal A used there is in S_β.

REGION 4. Right hereditary: This is dual to region 3.

REGION 5. Left and right hereditary: If F is a field other than GF(p), where p is prime, then the lower radical generated by F and β belongs here. It is not strongly hereditary since F will contain a subring (either Z or GF(p), depending on the characteristic of F) which is not radical. As shown in [10] it is not left or right strong.

REGION 6. Strongly hereditary: The lower radical generated by GF(p) and β is strongly hereditary. As above it is not left or right strong.

REGION 9. Left and right strong: Let η denote the idempotent radical. Let $\alpha = \ell(\beta \cup \eta)$. Then $S_\alpha = S_\beta \cap S_\eta$ [11]. Let $R \in S_\alpha$ and let a left ideal L of R be in α. Let $\beta(L) = A$. Then $A^2 \subseteq L\beta(L) \subseteq \beta(R) = 0$. Let $\eta(L/A) = B/A$. Then $B = B^2 + A$ and so $B^2 = (B^2 + A)^2 \subseteq B^3$. Hence $B^2 \in \eta$. η is the upper radical generated by all nilpotent rings and so is strict. $R \in S_\eta$ then implies $B^2 = 0$ and so $B = A$. Thus $L/A \in S_\eta \cap S_\beta = S_\alpha$. $L \in \alpha$ then implies $L = A$ and so $L \in \beta$. Since $R \in S_\beta$ and β is left strong we have $L = 0$. Thus α is left strong. Similarly α is right strong. The ring

$\begin{bmatrix} 2Z & 2Z \\ 2Z & 2Z \end{bmatrix}$ is in S_α but its left ideal $\begin{bmatrix} 2Z & 0 \\ 2Z & 0 \end{bmatrix}$ is not. Therefore α is not left stable and, similarly, α is not right stable. $Z \in \eta \subseteq \alpha$, while $2Z \in S_\alpha$. Therefore α is not hereditary.

REGION 10. Hereditary and left strong: Puczyłowski [7] has shown that the left strongly prime radical has these properties and that it is not right strong. Since it is left strong but not right strong it cannot be left or right hereditary. He uses a ring R consisting of all infinite matrices each with only finitely many nonzero columns over a certain simple domain. R is semisimple with respect to this radical but the left ideal L or R consisting of matrices all of whose columns are zero except the first is not semisimple as it contains a nonzero nilpotent ideal. Thus the radical is not left stable.

<u>REGION 7. Left strong</u>: Puczyłowski [7] modifies the above example for region 10 by taking the lower left strong radical generated by the left strongly prime radical and a polynomial ring F[x] over a certain field F. He shows that this radical is neither hereditary nor right strong. He shows that the infinite matrix R used above is still semisimple. So the same example shows that this radical is not left stable.

<u>REGION 8. Right strong</u>: This is dual to region 7.

<u>REGION 11. Hereditary and right strong:</u> This is dual to region 10. (Puczyłowski actually works on the right so his examples lie in regions 8, 11 and their duals in regions 7, 10.)

<u>REGION 12. Hereditary and left and right strong</u>: Let J denote the Jacobson radical and N_g the generalized nil radical. N_g is the upper radical generated by domains and so is strict. Each radical has the above properties and so $J \cap N_g$ also has these properties. Let R be a domain in J, e.g. R = $\{m/n \in Q$, where m is an even integer and n is odd$\}$. Then $\begin{bmatrix} R & R \\ R & R \end{bmatrix}$ belongs to J and also to N_g, since any nonzero homomorphic image contains divisors of zero. The left ideal $\begin{bmatrix} R & 0 \\ R & 0 \end{bmatrix}$ does not belong to N_g, since the domain R is an image. Therefore $J \cap N_g$ is not left hereditary. Similarly it is not right hereditary The standard example of 2×2 matrices over a field shows that $J \cap N_g$ is neither left nor right stable.

<u>REGION 13. Left and right hereditary and left and right strong</u>: The Jacobson radical belongs here.

<u>REGION 14. Strongly hereditary and left and right strong</u>: The lower Baer radical belongs here.

<u>REGION 17. Left and right stable</u>: Let α be the upper radical generated by all accessible subrings of $x\mathbb{R}[x]$. Since the ring is commutative, this set of rings is closed under formation of left and right ideals. Thus α is left

and right stable. Any nonzero accessible subring of $x\mathbb{R}[x]$ is uncountable and so cannot be a homomorphic image of the subring $xZ[x]$. Therefore $xZ[x]$ is radical and so α is not strict. Any homomorphic image of $\mathbb{R}[x]$ is idempotent and so cannot be isomorphic to a nonzero subring of $x\mathbb{R}[x]$. Therefore $\mathbb{R}[x]$ is radical, but contains the semisimple ideal $x\mathbb{R}[x]$. Thus α is not hereditary.

REGION 18. Strict: Let F be a field and let α be the upper radical generated by all subrings of $xF[x]$. Then α is strict. As above the ring $F[x]$ is radical and its ideal $xF[x]$ is semisimple. Thus α is not hereditary.

REGION 19. Hereditary and left stable: Beidar [2] gives an example of a radical, say ρ, which is hereditary and left stable. He shows that it is not right strong. Therefore it cannot be left or right hereditary. He uses a simple domain D which is not right noetherian. D may be defined as an algebra over any ground field F. ρ is the upper radical generated by all subdirectly irreducible rings with heart isomorphic to a left ideal of D. (Beidar's paper has been translated from Russian and the word translated as "strict" ought to have been translated as "strong".)

REGION 20. Hereditary and right stable: This is dual to region 19.

REGION 15. Left stable: Let $\sigma = \alpha \cap \rho$ where α is the radical defined for region 17 and ρ the radical defined for region 19. Since both α and ρ are left stable, σ is left stable [6]. We may assume that the field F and so the domain D used to define ρ has prime characteristic. Beidar shows that ρ is not right strong using a right ideal M of D showing that $M \in \rho$ while $D \in S_\rho$. Since $S_\rho \subseteq S_\sigma$ we have $D \in S_\sigma$. Since M has prime characteristic, no homomorphic image of M is a nonzero subring of $x\mathbb{R}[x]$. Thus $M \in \alpha$ and so $M \in \sigma$. It follows that σ is not right strong. Since $\mathbb{R}$ is divisible, no nonzero image of $\mathbb{R}[x]$ has prime characteristic. It follows that $\mathbb{R}[x] \in \rho$ and so that $\mathbb{R}[x] \in \sigma$. Since $x\mathbb{R}[x] \in S_\alpha \subset S_\sigma$ it follows that σ is not hereditary.

REGION 16. Right stable: This is dual to region 15.

REGION 21. Hereditary and left and right stable: Let F be a field which is not of prime order. Then the upper radical generated by F has these properties. Since F is not of prime order, F contains a radical subring and so this radical is not strict. The standard example of 2×2 matrices over F shows that it is neither left nor right hereditary.

REGION 22. Hereditary and strict: Let F be a field of prime order. Then the upper radical generated by F has these properties. As above it is not left or right hereditary.

REGION 23. Strongly hereditary and strict: As stated earlier this region consists of all hereditary A-radicals. The only hereditary A-radical containing β is the improper radical A consisting of all rings. Hereditary A-radicals have been classified. For each set P of primes there is a radical τ_P consisting of all rings R such that $(R,+)$ is the direct sum of its p-components, $p \in P$. These, together with the zero radical ω and the improper radical A, constitute all such radicals.

3. RADICALS CONTAINED IN THE LOWER BAER RADICAL

We have shown that the 23 regions in the Venn diagram are nonempty not just for radicals in general but also for radicals containing the lower Baer radical β. It seems natural to turn now to radicals contained in β. Here the situation for hereditary radicals is quite different. It has been shown by Armendariz [1] that a hereditary radical contained in β is the intersection of β with a hereditary A-radical. It follows that such radicals are strongly hereditary and so all the hereditary conditions coincide for radicals contained in β. Also, since each of these radicals is left and right strong, hereditary radicals contained in β are left and right strong. Thus regions 2,3,4,5,6,10,11,12,13,19,20,21,22 in the Venn diagram become empty when restricting attention to these radicals. We shall give examples to show that the remaining regions are nonempty. The list of properties and nonempty regions now becomes:

Strongly hereditary $\equiv$ {14,23};

Left strong $\equiv$ {7,9,14,15,17,18,23};

Right strong $\equiv$ {8,9,14,16,17,18,23};

Left stable $\equiv$ {15,17,18,23};

Right stable $\equiv$ {16,17,18,23};

Strict $\equiv$ {18,23}.

The following result is useful for constructing radicals with prescribed properties.

LEMMA: If R is a commutative and idempotent ring then the matrix ring $\begin{bmatrix} R & 0 \\ R & 0 \end{bmatrix}$ contains no nonzero left accessible subring which is a homomorphic image of R.

PROOF: If possible let A be such a subring. Then A is commutative and idempotent. An idempotent left accessible subring is a left ideal. Since $A^4 \neq 0$ there exist elements $\begin{bmatrix} x_i & 0 \\ y_i & 0 \end{bmatrix}$ in A, $1 \leq i \leq 4$, whose product is not zero. Then $zx_2x_3x_4 \neq 0$ either for $z = x_1$ or for $z = y_1$. Then

$$\begin{bmatrix} 0 & 0 \\ z & 0 \end{bmatrix} \begin{bmatrix} x_2 & 0 \\ y_2 & 0 \end{bmatrix} \quad \text{and} \quad \begin{bmatrix} x_3 & 0 \\ 0 & 0 \end{bmatrix} \begin{bmatrix} x_4 & 0 \\ y_4 & 0 \end{bmatrix}$$

are in A. However, these two elements do not commute. □

REGION 1. No properties: Let A be a Zassenhaus algebra over a field F [14], and let $\alpha = \ell(A)$. Then $A \in \beta$ and so $\alpha \subseteq \beta$. Since $A^2 = A$ we have $\alpha \subseteq \eta$, the idempotent radical. A contains nonzero nilpotent ideals which cannot be in η and so α is not hereditary. A is commutative and idempotent. It follows from the lemma that $\begin{bmatrix} A & 0 \\ A & 0 \end{bmatrix}$ is in S_α, while its right ideal $\begin{bmatrix} A & 0 \\ 0 & 0 \end{bmatrix}$ is in α. Therefore α is not right strong. Dually α is not left strong.

REGION 7. Left strong: Let A be this Zassenhaus algebra and let C consist of all left accessible subrings of $\begin{bmatrix} A & 0 \\ A & 0 \end{bmatrix}$. Let $\alpha = \beta \cap u(C)$. Then, as β and $u(C)$ are left strong, α is left strong. It follows from the lemma that

$A \in u(C)$ and so that $A \in \alpha$. As above α is not right strong. If α were hereditary it would be left hereditary which, with left strong, would imply right strong. Let K be a field of finite characteristic not equal to the characteristic of F. Then $\begin{bmatrix} K & K \\ K & K \end{bmatrix} \in S_\beta \subseteq S_\alpha$ but the left ideal $\begin{bmatrix} K & 0 \\ K & 0 \end{bmatrix}$ is not in S_α, as its ideal $\begin{bmatrix} 0 & 0 \\ K & 0 \end{bmatrix}$ is in β and in $u(C)$ and so in α. It follows that α is not left stable.

<u>REGION 8. Right strong:</u> This is dual to region 7.

<u>REGION 9. Left and right strong:</u> Let δ be the divisible radical and let $\alpha = \beta \cap \delta$. Then $\beta \cap \delta$ is left and right strong. α is not hereditary since Q^+ is in α and its ideal Z^+ is not. The standard example of 2×2 matrices over Q shows that α is neither left nor right stable.

<u>REGION 14. Strongly hereditary and left and right strong:</u> β itself belongs here. By the results of Armendariz the radicals in this region are classified as the intersection of β with each nonzero hereditary A-radical.

<u>REGION 17. Left and right stable:</u> Let $\alpha = \beta \cap \eta$. Since β is hereditary we have $\alpha(R) = \eta(\beta(R))$, for each ring R. Let $R \in S_\alpha$. Then $\eta(\beta(R)) = 0$ and so, since η is strict, $\beta(R)$ contains no nonzero idempotent subring. Let L be a left ideal of R. Then $\alpha(L) = \alpha(L)^2 \subseteq L\ \beta(L) \subseteq \beta(R)$. Hence $\alpha(L) = 0$ and so $L \in S_\alpha$. Therefore α is left stable and, similarly, α is right stable. Let A be a Zassenhaus algebra over a field F. Then $A \in \alpha$ and, as for region 1 here, α cannot be hereditary. As a vector space over F, $(A,+)$ has a ring E of linear transformations. E is in S_β and so in S_α. A embeds as a subring of E, where with each $a \in A$ we associate the mapping $x \to ax$. It follows that α is not strict.

<u>REGION 15. Left stable:</u> Let the class C be defined as for region 7 and let $\alpha = u(C) \cap \beta \cap \eta$. Since $u(C)$ and $\beta \cap \eta$ are left stable, α is left stable [6]. The same example as for regions 1 and 7 shows that α is not right strong. If α were hereditary then it would be left hereditary and this, with left stable, would imply that α is right strong.

<u>REGION 16. Right stable:</u> This is dual to region 15.

REGION 18. Strict: We show that this region consists of the nonhereditary A-radicals contained in β and classify these.

Let α be a strict radical contained in β. Let $S \in S_\beta$. Then $\begin{bmatrix} S & S \\ S & S \end{bmatrix} \in S_\beta \subseteq S_\alpha$. Since α is left stable we have $\begin{bmatrix} S & 0 \\ S & 0 \end{bmatrix} \in S_\alpha$ and then $S^+ \cong \begin{bmatrix} 0 & 0 \\ S & 0 \end{bmatrix} \in S_\alpha$. Now let p be a prime and let R be any ring in α. If $pR \neq R$ then the abelian group $(R/pR,+)$ may be given a multiplication making it into a direct sum of fields GF(p). From above it follows that $(R/pR)^+ \in S_\alpha$. $(R/pR,+)$ is a vector space over GF(p) and so has a ring E of linear transformations. Then $E \in S_\beta \subseteq S_\alpha$. Let B/pR be the left annihilator of R/pR. As before, left multiplication embeds R/B as a subring of E. Since α is strict $R/B \in S_\alpha$. Together with $R \in \alpha$ this implies B = R. It follows that $R/pR \cong (R/pR)^+$. Then $R \in \alpha$ and $(R/pR)^+ \in S_\alpha$ imply pR = R. Therefore every ring R in α is divisible. Let T be the torsion subgroup of R. Let R/T = U. Then U is divisible and torsion-free and so (U,+) is a vector space over Q. As above, using the ring of linear transformations, we can see that U = 0. Hence R is a divisible torsion-ring and so $R \cong R^+$. Thus all rings in α are trivial. Let P be the set of primes p such that $Z(p^\infty)^+ \in \alpha$. Then $R \cong \bigoplus_{p \in P} \oplus Z(p^\infty)^+$ and conversely any such ring is in α. It follows that $\alpha = \delta \cap \tau_P$ and so that α is an A-radical. Conversely any A-radical contained in β is strict and so has this form.

Gardner has drawn my attention to a preprint by Stewart and himself [5] in which they show that $\delta \cap \tau$ is the largest strict radical contained in the Jacobson radical J. Since $\beta \subseteq J$ this is a better result than that given above. An examination of the proof above shows that it will go through for a radical α if $S \in S_\alpha$ implies $\begin{bmatrix} S & S \\ S & S \end{bmatrix} \in S_\alpha$ and if the endomorphism rings of vector spaces over the fields GF(p) and Q belong to S_α. If S is the endomorphism ring of V then $\begin{bmatrix} S & S \\ S & S \end{bmatrix}$ is the endomorphism ring of $V \oplus V$. From this it is routine to check that the above results hold for the upper radical α generated by all these endomorphism rings.

REGION 23. Strongly hereditary and strict: The only hereditary A-radical contained in β is the zero radical ω, consisting only of the zero ring 0.

4. RADICALS CONTAINING THE IDEMPOTENT RADICAL

We consider now radicals containing the idempotent radical η. Easily we have that the only hereditary radical containing η is the improper radical A consisting of all rings. Any ring R can be embedded as an ideal in the idempotent ring (1,R). Then (1,R) being radical, together with the hereditary condition, implies that R is radical. Thus in this case in the Venn diagram regions 2,3,4,5,6,10,11,12,13,14,19,20,21,22 are empty and region 23 contains only the radical A. We now give examples showing that the remaining regions are nonempty.

<u>REGION 1. No properties</u>: Let α be the lower radical generated by η and the ring 2Z of even integers. α is not hereditary since $Z \in \alpha$ and $3Z \in S_\alpha$. It may be shown in a similar manner as previously that $\begin{bmatrix} 2Z & 2Z \\ 0 & 0 \end{bmatrix} \in S_\alpha$ while the left ideal $\begin{bmatrix} 2Z & 0 \\ 0 & 0 \end{bmatrix} \in \alpha$. Thus α is not left strong. Similarly α is not right strong.

<u>REGION 9. Left and right strong</u>: It has already been shown that $\ell(\beta \cup \eta)$ belongs here.

<u>REGION 7. Left strong</u>: Let A be a Zassenhaus algebra with coefficients from 2Z. Let C denote all left accessible subrings of $\begin{bmatrix} A & 0 \\ A & 0 \end{bmatrix}$. Let $\alpha = u(C) \cap \ell(\beta \cup \eta)$. Since each of these radicals is left strong it follows that α is left strong. α is not right strong since $\begin{bmatrix} A & 0 \\ A & 0 \end{bmatrix}$ is semisimple while its right ideal $\begin{bmatrix} A & 0 \\ 0 & 0 \end{bmatrix}$ is radical. That α is not left stable follows as for the example in region 9 in section 2. Clearly $\alpha \neq A$ and so α is not hereditary.

<u>REGION 8. Right strong</u>: This is dual to region 7.

<u>REGION 15. Left stable</u>: Let C consists of all left accessible subrings of $\begin{bmatrix} 2Z & 0 \\ 2Z & 0 \end{bmatrix}$ and $\alpha = u(C)$. Then α is left stable. α is not right strong since $\begin{bmatrix} 2Z & 0 \\ 2Z & 0 \end{bmatrix}$ is semisimple and it may be shown that its right ideal $\begin{bmatrix} 2Z & 0 \\ 0 & 0 \end{bmatrix}$ is radical. Again $\alpha \neq A$ and so α is not hereditary.

<u>REGION 16. Right stable</u>: This is dual to region 15.

<u>REGION 17. Left and right stable</u>: Let C consist of all accessible subrings

of $2Z[x]$ and $\alpha = u(C)$. Since $2Z[x]$ is commutative it follows that α is left and right stable. As above α is not hereditary. α is not strict since $2Z[x]$ is semisimple but it may be shown that no nonzero image of $2Z$ is an accessible subring of $2Z[x]$ and so that $2Z$ is a radical subring.

REGION 18. Strict: η itself belongs here.

5. RADICALS CONTAINED IN THE IDEMPOTENT RADICAL

THEOREM: A left strong radical γ contained in η is left stable. The only left strong hereditary radical contained in η is the zero radical ω.

PROOF: Let γ be a left strong radical contained in η. Let L be a left ideal of $R \in S_\gamma$. Since $\gamma \subseteq \eta$, $\gamma(L)$ is idempotent. It follows that $\gamma(I)$ is a left ideal of R. Since γ is left strong we have $\gamma(L) = 0$ and so $L \in S_\gamma$. Therefore γ is left stable.

Now let γ also be hereditary and let $R \in \gamma$. The left ideals $\begin{bmatrix} R & 0 \\ 0 & 0 \end{bmatrix}$ and $\{\begin{bmatrix} r & r \\ 0 & 0 \end{bmatrix} \; r \in R\}$ of the ring $\begin{bmatrix} R & R \\ 0 & 0 \end{bmatrix}$ are each isomorphic to R and so are in γ. Since γ is left strong and $\begin{bmatrix} R & R \\ 0 & 0 \end{bmatrix}$ is their sum it follows that $\begin{bmatrix} R & R \\ 0 & 0 \end{bmatrix} \in \gamma$. Since γ is hereditary it follows that $\begin{bmatrix} 0 & R \\ 0 & 0 \end{bmatrix} \in \gamma$. Then $\gamma \subseteq \eta$ implies $\begin{bmatrix} 0 & R \\ 0 & 0 \end{bmatrix} = \begin{bmatrix} 0 & R \\ 0 & 0 \end{bmatrix}^2 = \begin{bmatrix} 0 & 0 \\ 0 & 0 \end{bmatrix}$, and so $R = 0$. Thus $\gamma = \omega$, as required. □

From this theorem and its dual it follows that the middle regions 7,8,9,10,11,12,13,14 of the Venn diagram are empty, as are the regions 19,20,21,22, while the region 23 contains only the zero radical ω. However, we have an unsolved problem in this case. The previous examples separating left hereditary and right hereditary are not contained in η and we have not been able to determine whether regions 3 and 4 are empty or nonempty, i.e. there remains the problem, for radicals in which every ring is idempotent, as to whether or not left hereditary implies right hereditary. For the remaining regions 1,2,5,6,15,16,17,18 we list examples showing that these are not empty. The examples have either been given before or are very similar to ones given before so we omit the details of proofs.

REGION 1. No properties: $\ell(Z)$ belongs here.

REGION 2. Hereditary: $\ell\begin{bmatrix} F & F \\ F & F \end{bmatrix}$ belongs here, where F is any field.

REGION 5. Left and right hereditary: $\ell(F)$ belongs here for any field F not of prime order.

REGION 6. Strongly hereditary: $\ell(GF(p))$ belongs here for any prime p.

REGION 15. Left stable: Let C denote all left accessible subrings of $\begin{bmatrix} F & 0 \\ F & 0 \end{bmatrix}$ where F is a field, and N all nilpotent rings. Then $\alpha = u(C \cup N)$ belongs here.

REGION 16. Right stable: This is dual to region 15.

REGION 17. Left and right stable: Let F be any field not of prime order. Then $\alpha = u(F,N)$ belongs here.

REGION 18. Strict: η belongs here.

ACKNOWLEDGEMENT

Thanks are due to the Royal Society of Edinburgh for financial support helping me to attend this conference.

REFERENCES

[1] E.P. Armendariz, Hereditary subradicals of the lower Baer radical, Publ. Math. Debrecen 15,(1968) 91-93.

[2] K.I. Beidar, Examples of rings and radicals, Colloq. Math. Soc. Bolyai, 38, Radical Theory, Eger, 1982, North-Holland, Amsterdam, 1985, 19-46.

[3] N.J. Divinsky, Rings and Radicals, Toronto, 1965.

[4] B.J. Gardner, Radicals of abelian groups and associative rings, Acta Math. Acad. Sci. Hung. 29, (1973) 259-268.

[5] B.J. Gardner and P.N. Stewart, The survival of the Jacobson radical in some ring extensions, Preprint.

[6] E.R. Puczyłowski, Remarks on stable radicals, Bull. Acad. Polon. Sci. 28 (1980), 11-16.

[7] E.R. Puczyłowski, On Sands' questions concerning strong and stable radicals, Glasgow Math. J. 28 (1986), 1-3.
[8] E.R. Puczyłowski, On questions concerning strong radicals of associative rings, Preprint.
[9] A.D. Sands, Strong upper radicals, J. Math., Oxford (2), 27 (1976), 21-24.
[10] A.D. Sands, On relations among radical properties, Glasgow Math. J. 18 (1977), 17-23.
[11] R.L. Snider, Lattices on radicals, Pacific J. Math. 40 (1972), 207-220.
[12] P.N. Stewart and R. Wiegandt, Quasi-ideals and bi-ideals in radical theory, Acta Math. Acad. Sci. Hung. 39 (1982), 289-294.
[13] R. Tangeman and D. Kreiling, Lower radicals in non-associative rings, J. Austral. Math. Soc. 14 (1972), 419-423.
[14] R. Wiegandt, Radical and semi-simple classes of rings, Kingston, Ontario, 1974.

A.D. Sands
Department of Mathematics and
Computer Science,
The University
Dundee DD1 4HN
U.K.

STEFAN VELDSMAN

Sufficient conditions for a well-behaved Kurosh-Amitsur radical theory II

Recently three approaches were given (independently) by Puczyłowski [3], Beidar [1] and the author [4] to formulate conditions on a universal class of "algebras" to ensure that the following important questions in general radical theory have positive answers: Are semisimple classes hereditary? Do radical classes have the ADS-property (i.e. $\mathcal{R}(I) \triangleleft A$ for $I \triangleleft A$)? Can semisimple classes be characterized by closure conditions (e.g. is semisimple = coradical)? Does Sands' theorem hold (i.e. is semisimple = regular, coinductive and closed under extensions)? And lastly, does the lower radical construction terminate?

Puczyłowski [3] showed that, in what he called a *normal class*, the above questions all have positive answers. In [4] a sequence of conditions, viz. the $d_i(A)$-classes, $i = 0,1,2,3,4$, were defined in which some or all the above questions have positive answers. The normal classes "essentially" coincide with the $d_3(\mathcal{B})$-classes. Beidar [1] obtained the same results as in [3] and [4] in a universal class of R-algebras (R associative, commutative, $1 \in R$), but from a different set of starting axioms. It is our purpose here to show how the conditions of Beidar fit into the scheme of $d_i(A)$-classes since we know from [4] that any universal class in which all radicals have the ADS-property must be a d_1-class.

1. PRELIMINARIES

Let $\mathcal{W}$ be a universal class of Ω-groups which is closed under the formation of finite direct sums (we need this in the proof of Lemma 2.1). All considerations will be in $\mathcal{W}$. Let $A \subseteq \mathcal{W}$ be a fixed subclass of $\mathcal{W}$ with $A \setminus \{0\} \neq \emptyset$. Radical classes will be in the sense of Kurosh and Amitsur (cf. [3], [4], and their references for definitions and more information). As usual, $\mathcal{U}$ and S will denote the upper radical and semisimple operators respectively. Ideals and accessible Ω-subgroups will be denoted by $\triangleleft$ and $\triangleleft\triangleleft$ respectively. All chosen subclasses of $\mathcal{W}$ will be assumed to be abstract.

From [4] we recall that $\mathcal{W}$ is a $d_i(A)$-*class* if $J \triangleleft I \triangleleft A \in \mathcal{W}$ and J is not an ideal of A, then there exists a surjective homomorphism $\delta: B \to C/J$ with:

For $i = 0$: $B \triangleleft \triangleleft J$, $J \subsetneqq C \triangleleft \triangleleft I$, say $C/J = C_1/J \triangleleft C_2/J \triangleleft \ldots \triangleleft C_{m-1}/J \triangleleft C_m/J = I/J$ and $C_{m-1}/J \in \mathcal{A}$.

For $i = 1$: As for $i = 0$ but with $B = J$.

For $i = 2$: $B = J$, $C \triangleleft I$, $0 \neq C/J \in \mathcal{A}$.

For $i = 3$: $B = J$, $C \triangleleft I$, $0 \neq C/J \in \mathcal{A}$ and $\ker \delta \triangleleft I$.

We will shortly use d_i-*classes* for $d_i(\mathcal{W})$-classes.

As in Higgins [2], an Ω-group A is *solvable* if $A^{(n)} = 0$ for some $n \geq 0$ where $A^{(n)}$ is defined inductively by $A^{(0)} = A$, $A^{(n)} = [A^{(n-1)}, A^{(n-1)}]$. If A is solvable, the smallest n for which $A^{(n)} = 0$ is the *degree of solvability* of A.

By $\mathcal{S}$ we will denote the class $\mathcal{S} = \{A \in \mathcal{W} \mid A \text{ is solvable}\}$.

ON $d_3(\mathcal{S})$-CLASSES

From [4] we know that if $\mathcal{W}$ is a $d_3(\mathcal{A})$-class, then all semisimple classes are hereditary, every radical class has the ADS-property and the semisimple classes $\mathcal{M}$ are characterized as those classes which are regular, coinductive, closed under extensions and satisfying

$$I \triangleleft A \in \mathcal{M} \text{ and } I \in \mathcal{A} \quad \text{imply} \quad I \in \mathcal{M}. \qquad (*)$$

For certain choices of $\mathcal{A}$, condition $(*)$ is a consequence of the other conditions imposed on $\mathcal{M}$, e.g. if $\mathcal{A}$ is

$$\mathcal{B} = \{A \in \mathcal{W} \mid J \triangleleft I \triangleleft A \text{ implies } J \triangleleft A\}$$

$$\mathcal{T} = \{A \in \mathcal{W} \mid A^{(1)} = 0\}$$

and if the underlying groups in $\mathcal{W}$ are abelian,

$$\mathcal{Z} = \{A \in \mathcal{W} \mid A\Omega = 0\}$$

would suffice where $A\Omega = \{\underline{a}\omega \mid \underline{a} \in A, \omega \in \Omega\}$. If $\mathcal{A}$ is the class of all nilpotent Ω-groups (cf. Higgins [2]) in $\mathcal{W}$, we have to impose further restrictions on $\mathcal{W}$ to have the same conclusions; cf. Theorem 3.19 in [4]. However, using a method developed by Beidar, the same conclusions can be made in the more general $d_3(\mathcal{S})$-classes.

Consider the following condition that the class of solvable Ω-groups $\mathcal{S}$ may satisfy

(S1) If $B \lhd A$, $A^{(n)} = 0$, $(A/B)^{(1)} = 0$ and $B^{(n-1)} \neq 0$, then there exists a homomorphism $f:A \to A^{(n-1)}$ with $f(B) \neq 0$.

LEMMA 2.1: Suppose $\mathcal{S}$ satisfies condition (S1) in $\mathcal{W}$. Let $\mathcal{M} \subset \mathcal{W}$ be regular, coinductive, closed under extensions and with $\mathcal{SUM}$ hereditary. If $A \in \mathcal{SUM}$ and $A^{(n)} = 0$, then $A \in \mathcal{M}$.

PROOF: (cf. Beidar [1], Corollary 1.4): Let $\mathcal{R} = \mathcal{UM}$ and we now proceed by induction on n. For n = 1, it follows from Lemma 3.10 in [4]. Suppose thus $n \geq 2$. Since $A^{(1)} \subseteq A$, $A/A = 0 \in \mathcal{M}$ and $\mathcal{M}$ is coinductive, choose $B \lhd A$ minimal with respect to $A^{(1)} \subseteq B$ and $A/B \in \mathcal{M}$.

Suppose $B^{(n-1)} \neq 0$. By (S1) there exists a homomorphism $f:A \to A^{(n-1)}$ with $f(B) \neq 0$. Since $(A^{(n-1)})^{(1)} = A^{(n)} = 0$, we have $f(A) \lhd A^{(n-1)}$. Thus

$$f(A) \lhd A^{(n-1)} \lhd A^{(n-2)} \lhd \dots \lhd A^{(1)} \lhd A^{(0)} = A \in \mathcal{SR}$$

and from the hereditariness of $\mathcal{SR}$ we have $f(A) \in \mathcal{SR}$. But $(f(A))^{(1)} = 0$ and from the case n = 1, $f(A) \in \mathcal{M}$.

Let $\bar{A} = A/B$ and let $D = \bar{A} \oplus f(A)$. Define $h:A \to D$ by

$$h(x) = (x + B, f(x)), \quad x \in A.$$

Then h is a homomorphism and $f(A) \neq 0$ implies $h(A) \neq 0$. From $A^{(1)} \subseteq B$ and $f(A^{(1)}) = f(A)^{(1)} \subset (A^{(n-1)})^{(1)} = 0$, $h(A^{(1)}) = 0$ and $A^{(1)} \subseteq \ker h$ follows. Note also that $\ker h \subsetneq B$ and $D^{(1)} = 0$. Using $\bar{A} \in \mathcal{M}$, $D/\bar{A} \cong f(A) \in \mathcal{M}$ and the fact that $\mathcal{M}$ is closed under extensions, we have $D \in \mathcal{M}$. From $h(A) \lhd D \in \mathcal{M} \subset \mathcal{SR}$ it follows that $h(A) \in \mathcal{SR}$. Since $h(A)^{(1)} = 0$, we have from the case n = 1, $h(A) \in \mathcal{M}$. Thus we have $A/\ker h \cong h(A) \in \mathcal{M}$, $A^{(1)} \subseteq \ker h$ and $\ker h \subset B$, which contradicts the choice of B. Thus $B^{(n-1)} = 0$ and since $B \lhd A \in \mathcal{SR}$ and $\mathcal{SR}$ is hereditary, the induction assumption yields $B \in \mathcal{M}$. But also $A/B \in \mathcal{M}$; hence $A \in \mathcal{M}$ which completes the proof. □

Combining Lemma 2.1 and Theorem 3.2 from [4], we have:

THEOREM 2.2: Let $\mathcal{W}$ be a $d_3(\mathcal{S})$-class in which $\mathcal{S}$ satisfies condition (S1). Then a subclass of $\mathcal{W}$ is a semisimple class if and only if it is regular, coinductive and closed under extensions.

3. ON $d_1(\mathcal{S})$-CLASSES

Any universal class $\mathcal{W}$ in which all the radical classes have the ADS-property must be a d_1-class. Conversely, if $\mathcal{W}$ is a d_1-class in which all the semisimple classes are hereditary, then every radical class in $\mathcal{W}$ has the ADS-property (cf. [4]). The converse also holds in certain $d_1(\mathcal{S})$-classes. A chain $C_1 \lhd C_2 \lhd \ldots \lhd C_m \in \mathcal{W}$ has the *descending degree of solvability* (dds for short) if, whenever C_k is solvable, then the degree of solvability of C_{k-1} is strictly less than that of C_k, $k \in \{2,3,\ldots,m\}$.

$\mathcal{W}$ is a *$d_1(\mathcal{S})$-class with dds* if $J \lhd I \lhd A \in \mathcal{W}$ and J not an ideal of A implies the existence of a surjective homomorphism $\delta: J \to C/J$ with $C \lhd \lhd I$, say

$$0 \neq C/J = C_1/J \lhd C_2/J \lhd \ldots \lhd C_{m-1}/J \lhd C_m/J = I/J$$

with $C_{m-1}/J \in \mathcal{S}$ and the chain

$$C/J = C_1/J \lhd C_2/J \lhd \ldots \lhd C_{m-1}/J \lhd C_m/J = I/J$$

has the dds.

In general, the degree of solvability of C_{m-1}/J depends on the chain $J \lhd I \lhd A$. If, however, it is independent of the choice of I and J but only depends on A, we call $\mathcal{W}$ a *$d_1(\mathcal{S})$-class with dds and a fixed degree of solvability.*

THEOREM 3.1: Let $\mathcal{W}$ be a $d_1(\mathcal{S})$-class with dds. Then every radical class in $\mathcal{W}$ has the ADS-property.

PROOF: Let $\mathcal{R} \subseteq \mathcal{W}$ be a radical class in $\mathcal{W}$ and consider $B \lhd A \in \mathcal{W}$. In order to show that $\mathcal{R}(B) \lhd A$, we divide the proof into the following three steps:

(1) If $B \in \mathcal{S}$ then $\mathcal{R}(B) \triangleleft A$.

(2) If $B \in \mathcal{S}$ and $A \in S\mathcal{R}$, then $J \in S\mathcal{R}$ for all accessible Ω-subgroups J of B

(3) $\mathcal{R}(B) \triangleleft A$.

(1) Let $B \triangleleft A$ with $B^{(n)} = 0$. We proceed by induction on n. Suppose $n = 1$ and suppose $J := \mathcal{R}(B)$ is not an ideal of A. Then there exists a surjective homomorphism

$$\delta: J \to C/J = C_1/J \triangleleft C_2/J \triangleleft \ldots \triangleleft C_{m-1}/J \triangleleft C_m/J = B/J.$$

Since $B^{(1)} = 0$, so is $(B/J)^{(1)} = 0$. Hence $0 \neq C/J \triangleleft B/J \in S\mathcal{R}$, which contradicts $C/J \in \mathcal{R}$. Thus $\mathcal{R}(B) \triangleleft A$.

Suppose $n \geqq 2$ and $J := \mathcal{R}(B)$ is not an ideal of A. Then there is a surjective homomorphism

$$\delta: J \to C/J = C_1/J \triangleleft C_2/J \triangleleft \ldots \triangleleft C_{m-1}/J \triangleleft C_m/J = B/J$$

and this chain has the dds. Since $(B/J)^{(n)} = 0$, the degree of solvability of B/J is k for some $k \leqq n$. If $k = 1$, then $0 \neq C/J \in \mathcal{R}$, which contradicts $C/J \triangleleft B/J \in S\mathcal{R}$. Thus $k \geqq 2$. Assume the degree of solvability of C_{m-1}/J is q. By the dds, $q < k$. From the induction assumption on n, we have $\mathcal{R}(C_{m-1}/J) \triangleleft B/J \in S\mathcal{R}$; hence $C_{m-2}/J \in S\mathcal{R}$.

Now C_{m-2}/J has degree of solvability say p, with $p < q < k \leqq n$, and as above we get $C_{m-2}/J \in S\mathcal{R}$. Continuing in this way we get $0 \neq C/J \in \mathcal{R} \cap S\mathcal{R} = 0$ clearly impossible.

(2) Consider $J = J_1 \triangleleft J_2 \triangleleft \ldots \triangleleft J_{m-1} \triangleleft J_m = B \triangleleft A \in S\mathcal{R}$ with $B^{(n)} = 0$. From the first part $\mathcal{R}(J_{m-1}) \triangleleft J_m \in S\mathcal{R}$; hence $J_{m-1} \in S\mathcal{R}$. Continuing in this way we get $J \in S\mathcal{R}$.

(3) This is now obvious from the above, which completes the proof. □

Recall from the definition of a $d_3(A)$-class and the proof of Theorem 3.2 in [4] the fact that ker $\delta \triangleleft I$ played an important role. Once again, using a technique of Beidar this can be generalized. For ease of reference we have to introduce another type of class:

$\mathcal{W}$ is a $d_3^*(\mathcal{S})$-*class* if $J \triangleleft I \triangleleft A \in \mathcal{W}$ and J is not an ideal of A, then there exists a surjective homomorphism

$$\delta : J \to C/J = C_1/J \triangleleft C_2/J \triangleleft \ldots \triangleleft C_{m-1}/J \triangleleft C_m/J = I/J$$

and an ideal $L \triangleleft I$ such that $L \subseteq \ker \delta$, $J/L \in \mathcal{S}$, $C_{m-1}/J \in \mathcal{S}$, $C/J \neq 0$ and the chain

$$C/J = C_1/J \triangleleft C_2/J \triangleleft \ldots \triangleleft C_{m-1}/J \triangleleft C_m/J = I/J$$

has the dds.

<u>THEOREM 3.2</u>: Let $\mathcal{W}$ be a $d_3^*(\mathcal{S})$-class in which $\mathcal{S}$ satisfies condition (S1). Then a subclass of $\mathcal{W}$ is a semisimple class if and only if it is regular, coinductive and closed under extensions.

<u>PROOF</u> (cf. Beidar [1], Theorem 1.1): Only the necessity needs verification. Choose $\mathcal{M} \subseteq \mathcal{W}$ satisfying the conditions and let $\mathcal{R} = \mathcal{U}\mathcal{M}$. Let $A \in S\mathcal{R}$; we show $A \in \mathcal{M}$. Firstly note that if $X \in S\mathcal{R}$ and X is solvable, then $X \in \mathcal{M}$ since $S\mathcal{R}$ is hereditary and from Lemma 2.1. Choose $I \triangleleft A$ minimal with respect to $0 \neq A/I \in \mathcal{M}$. If $I = 0$, we are done. Suppose thus $I \neq 0$. Choose $J \triangleleft I$ minimal with respect to $0 \neq I/J \in \mathcal{M}$. If $J \triangleleft A$, then $A/J \in \mathcal{M}$ holds since $\mathcal{M}$ is closed under extensions, which contradicts the choice of I. Thus J is not an ideal of A and consequently there is a surjective homomorphism

$$\delta : J \to C/J = C_1/J \triangleleft C_2/J \triangleleft \ldots \triangleleft C_{m-1}/J \triangleleft C_m/J = I/J$$

and an ideal $L \triangleleft I$ with $L \subseteq \ker \delta$ and $J/L \in \mathcal{S}$. If $J/L \in \mathcal{R}$, then $0 \neq C/J \cong J/\ker \delta \in \mathcal{R}$. But $C/J \triangleleft \triangleleft I/J \in S\mathcal{R}$ and $S\mathcal{R}$ is hereditary - clearly impossible. Hence J/L is not in $\mathcal{R}$. Now $J/L \triangleleft I/L$ is minimal with respect to $(I/L)/(J/L) \cong I/J \in \mathcal{M}$. Since J/L is solvable, so is $(J/L)/\mathcal{R}(J/L) \in S\mathcal{R}$. Thus it is in $\mathcal{M}$. From Theorem 3.1, $\mathcal{R}(J/L) \triangleleft I/L$ and since $\mathcal{M}$ is closed under extensions $(I/L)/\mathcal{R}(J/L) \in \mathcal{M}$. But J/L is minimal with respect to $(I/L)/(J/L) \in \mathcal{M}$. Thus $J/L = \mathcal{R}(J/L)$ - a contradiction. Hence $I = 0$, which completes the proof. □

4. THE TERMINATION OF THE LOWER RADICAL

Let $M \subseteq W$. The Kurosh chain for M can be defined as follows:

M_1 is the homomorphic closure of M and if M_β has been defined for $\beta < \alpha \in \text{Ord}$, let $M_\alpha := \{A \in W \mid$ every nonzero homomorphic image of A has a nonzero ideal which is in M_β for some $\beta < \alpha\}$

The *lower radical determined by* M, denoted by $1M$, is defined by

$$1M := \bigcup_\alpha M_\alpha.$$

LEMMA 4.1: Let W be a $d_1(S)$-class with the dds. Suppose S satisfies:

(S2) $A \in S$ and $A^{(1)} \neq 0$ implies the existence of a nonzero surjective homomorphism $h:A \to B \triangleleft \triangleleft A^{(1)}$.

If R is a radical class and $A \in R \cap S$, then $A^{(1)} \in R$.

PROOF: By the ADS-property, $R(A)^{(1)}) \triangleleft A$. Let $C = A^{(1)}/R(A^{(1)})$ and let $D = A/R(A)$. Assume $C \neq 0$. Then $D^{(1)} \neq 0$ and since A is solvable, so is D. From (S2) there is a nonzero homomorphism $h:D \to B \triangleleft \triangleleft D^{(1)}$, $B = h(D) \neq 0$. Then $B \in R$. But SR is hereditary and $D^{(1)} \neq C \in SR$ - a contradiction. Hence $C = 0$ and $A^{(1)} = R(A^{(1)}) \in R$. □

LEMMA 4.2: Let W and S be as in the previous lemma. Let M be a homomorphical closed subclass of W and let $R := 1M$. If $D \in R$ and $D^{(n)} = 0$, then $D \in M_{n+1}$.

PROOF: (by induction on n): For $n = 1$, consider the nonzero homomorphic image B of D. Then $B \in R$; hence there is $0 \neq I \triangleleft \triangleleft B$ with $I \in M_1$. But $B^{(1)} = 0$, so $I \triangleleft B$. Thus $D \in M_2$. Assume the result holds for all k, $k < n$ $(n > 1)$. Using the above lemma, $D^{(1)} \in R$. Furthermore, $(D^{(1)})^{(n-1)} = D^{(n)} = 0$ and by the induction assumption $D^{(1)} \in M_{n-1+1} = M_n$.

Now $(D/D^{(1)})^{(1)} = 0$ and $(D/D^{(1)}) \in R$ and from the first case $(D/D^{(1)}) \in M_2$. Using this and the fact that $D^{(1)} \in M_n$, we conclude that $D \in M_{n+1}$. □

For our next result we need a certain construction. Let $A \in \mathcal{W}$ and let $\mathcal{M} \subseteq \mathcal{W}$ and let $\mathcal{M} \subseteq \mathcal{W}$ be a homomorphically closed class. Choose $n \geq 1$ fixed. Define a chain of ideals of A as follows:

> $A_1 = \sum (I \triangleleft A \mid I \in \mathcal{M}_n)$. Suppose A_β has been defined for $\beta < \alpha \in \mathrm{Ord}$. If α is not a limit ordinal, let A_α be the ideal of A for which $A_\alpha/A_{\alpha-1} = \sum [I/A_{\alpha-1} \mid I/A_{\alpha-1} \triangleleft A/A_{\alpha-1}$ and $I/A_{\alpha-1} \in \mathcal{M}_n]$. If α is a limit ordinal, let $A_\alpha = \cup (A_\beta \mid \beta < \alpha)$.

Let

$$\mathcal{M}_n(A) := \sum (A_\alpha \mid \alpha \in \mathrm{Ord}).$$

It can be verified that $\mathcal{M}_n(A) \in \mathcal{M}_{m+1}$ and that $A/\mathcal{M}_n(A)$ has no nonzero ideals in $\mathcal{M}_n$.

<u>THEOREM 4.3</u>: Let $\mathcal{W}$ be a $d_1(\mathcal{S})$-class with dds and a fixed degree of solvability and assume $\mathcal{S}$ satisfies (S2). Let $\mathcal{M}$ be a homomorphically closed subclass of $\mathcal{W}$. Then $1\mathcal{M} = \mathcal{M}_\omega$, i.e. the lower radical construction terminates at the first limit ordinal.

<u>PROOF</u>: Since $\mathcal{M}_{\omega+1}$ is homomorphically closed, it is sufficient to show that if $A \in \mathcal{M}_{\omega+1}$ then A has a nonzero ideal in $\mathcal{M}_q$ for some $q < \omega$. Because $A \in \mathcal{M}_{\omega+1}$ there exists a chain $0 \neq K \triangleleft I \triangleleft A$ with $K \in \mathcal{M}_t$ for some $t < \omega$. If $K \triangleleft A$, then we are done. Suppose thus K is not an ideal in A. By the assumption on $\mathcal{W}$ and by the definition, let n be the fixed degree of solvability obtained from the chain $K \triangleleft I \triangleleft A$ where K is not an ideal in A. Since $\mathcal{M}_t \subseteq (\mathcal{M}_t)_{n+1}$ we have $0 \neq K \subseteq J := (\mathcal{M}_t)_{n+1}(I) \triangleleft I \triangleleft A$. If J is not an ideal in A, there exists a surjective homomorphism

$$\delta: J \to C/J = C_1/J \triangleleft C_2/J \triangleleft \dots \triangleleft C_{m-1}/J \triangleleft C_m/J = I/J$$

with C_{m-1}/J solvable of degree n. Since $J \in (\mathcal{M}_t)_{n+2}$, $C/J \in (\mathcal{M}_t)_{n+2} \subseteq \mathcal{R}$. Let $X/J := \mathcal{R}(C_{m-1}/J) \triangleleft I/J$ (by the ADS). Since $\mathcal{SR}$ is hereditary and $C/J \in \mathcal{R}$, we have $0 \neq X/J$. But the degree of solvability of X/J is k, $k \leq n$;

thus from Lemma 4.2, $X/J \in M_{k+1} \subseteq M_{n+1} \subseteq (M_t)_{n+1}$. But I/J has no nonzero ideals in $(M_t)_{n+1}$. Hence $J \triangleleft A$ holds.

Lastly, $J = (M_t)_{n+1}(I) \in (M_t)_{n+2} \subseteq M_{t+n+2-1} = M_{t+n+1}$, which completes the proof. □

5. ON BEIDAR'S CONDITIONS

We now show that the conditions considered by Beidar for universal classes of algebras are sufficient to enforce classes of the type considered in the previous sections. In the sequel, U will denote a universal class of R-algebras (R associative, commutative, $1 \in R$).

The class U may satisfy the following conditions (cf. Beidar [1]):

(B1) To every $C \triangleleft B \triangleleft A \in U$ and $a \in A$, there exists an integer $m \geq 1$ such that $aC^{(m)} + C^{(m)}a \subseteq C$.

(B2) To every $C \triangleleft B \triangleleft A \in U$ and $a \in A$, there exists an integer $n \geq 1$ such that $(\langle aC + Ca + C\rangle_B)^{(n)} \subseteq C$ (where $\langle X\rangle_Y$ denotes the ideal in the algebra Y generated by the subset $X \subseteq Y$).

(B2)* The same as in (B2), except the integer n depends only on A and not on C and B.

(B3) To every $C \triangleleft B \triangleleft A \in U$ there exists an integer $t \geq 1$ such that $\langle C^{(t)}\rangle_B \subseteq C^{(1)}$.

As examples of classes satisfying these conditions, Beidar gave the varieties of associative algebras, alternative algebras, Jordan algebras with $1/2 \in R$, (γ,δ)-algebras with $1/6 \in R$, and Andrunakievic s-varieties.

From [1], it can be verified that the class of solvable algebras S in U satisfies conditions (S1) and (S2).

PROPOSITION 5.1: If U satisfies conditions (B1) and (B2), then U is a $d_1(S)$-class with the dds.

PROOF: Consider $J \triangleleft I \triangleleft A \in U$ with J not an ideal in A. Then there is an $a \in A$ such that $aJ + Ja \not\subseteq J$; assume $aJ \not\subseteq J$. Then

$$0 \neq K := \langle aJ + Ja + J\rangle_I/J \triangleleft I/J.$$

y (B2), $K^{(n)} = 0$ for some n. Define a function $f:J \to K$ by $f(x) = ax + J$, $\in J$. Then f is a nonzero R-module homomorphism and by (B1) $f(J^{(m)}) = 0$ or some m. From Lemma 1.1 in Beidar [1], there is a nonzero R-algebra omomorphism

$$h:J \to \mathrm{Ann}_{K^{(p)}}(K^{(p)})$$

or some $0 \leqq p \leqq n-2$. Then

$$h:J \to h(A) \triangleleft \mathrm{Ann}_{K^{(p)}}(K^{(p)}) \triangleleft K^{(p)} \triangleleft K^{(p-1)} \triangleleft \dots \triangleleft K^{(1)} \triangleleft K^{(0)} = K \triangleleft I/J$$

nd this chain has the dds.

ROPOSITION 5.2: Suppose $\mathcal{U}$ satisfies conditions (B1), (B2) and (B3). Then is $d_3^*(\mathcal{S})$-class.

ROOF: Consider $J \triangleleft I \triangleleft A \in \mathcal{U}$ with J not an ideal in A. As in the previous roof, we get a homomorphism and a chain with the dds

$$h:J \to h(A) \triangleleft \mathrm{Ann}_{K^{(p)}}(K^{(p)}) \triangleleft K^{(p)} \triangleleft K^{(p-1)} \triangleleft \dots \triangleleft K^{(1)} \triangleleft K^{(0)} = K \triangleleft I/J$$

here K is solvable. By (B3), there exists a $t \geqq 0$ such that

$$L := \langle J^{(t)}\rangle_I \subseteq J^{(1)}.$$

hen $L \triangleleft I$ and $L \subseteq \ker h$. □

The proof of the last result is as in Proposition 5.1.

ROPOSITION 5.3: If $\mathcal{U}$ satisfies conditions (B1) and (B2)*, then $\mathcal{U}$ is a $d_1(\mathcal{S})$-class with dds and a fixed degree of solvability.

REFERENCES

[1] K.I. Beidar, Semisimple classes of algebras and the lower radical, Mat. Issled. Kishinev, to appear.

[2] R.J. Higgins, Groups with multiple operators, London Math. Soc. Proc. 6 (1956), 366-416.

[3] E.R. Puczyłowski, On general theory of radicals, Preprint.

[4] S. Veldsman, Sufficient conditions for a well-behaved Kurosh-Amitsur radical theory, Proc. Edinburgh Math. Soc., to appear.

S. Veldsman
Department of Mathematics
University of Port Elizabeth
P.O. Box 1600
6000 Port Elizabeth
South Africa

RICHARD WIEGANDT

Recent results in the general radical theory of rings and ring-like structures

During the last couple of years successive attempts have been made to prove significant results in the Kurosh-Amitsur radical theory of associative algebras as well as in developing the Kurosh-Amitsur radical theory in other types of algebras, such as nonassociative algebras, alternative algebras, Jordan algebras, Lie algebras, (γ,δ)-algebras, Andrunakievich varieties, involution algebras, near-rings, topological rings, etc. In the most recent papers Beidar [14], Puczyłowski [50] and Veldsman [65] have imposed conditions on the algebras considered, and derived important results first of all concerning characterizations of semisimple classes and termination of Kurosh's lower radical construction. These results make it possible to develop a general theory for radicals of various types of algebras. The conditions of Beidar are purely conditions on not necessarily associative algebras, while the conditions used by Puczyłowski and Veldsman are modified versions of Terlikowska-Osłowska's [60] category theoretical condition. In organizing and presenting the material in this survey article, my point of view has been strongly influenced by the papers [14] and [50]. (Sincere thanks are due to K.I. Beidar and E.R. Puczyłowski for making me acquainted with the content of their papers [14] and [50] before publication.) During the Conference, to my surprise, S. Veldsman presented his paper [65] in which the investigations are closely related to those of Puczyłowski [50]; in fact, he dealt with various versions of Terlikowska-Osłowska's condition. Thus in the final version of this survey I have taken into consideration also the results of Veldsman, who kindly gave me a copy of [65]. Many of the above-mentioned types of algebras satisfy Beidar's condition, or Terlikowska-Osłowska's condition, recovering in this way known results as well as obtaining numerous new ones. We shall therefore concentrate only on the Anderson-Divinsky-Suliński (ADS) property of radicals, on characterizing semisimple classes and on the termination of Kurosh's lower radical construction. In proving purely radical theoretical results many interesting properties of algebras have been discovered (cf. [3], [13], [26], [57]) and we shall explicitly mention some of them at appropriate places.

1. PRELIMINARIES

Radical theory can be developed in any *universal class* (that is, a class closed under taking homomorphic images and ideals) of algebras, but for the sake of simplicity we shall consider here always *varieties* of algebras. Our attention is focused on associative and alternative algebras, Jordan algebras, involution algebras, and on Andrunakievich s-varieties of not necessarily associative algebras, but we shall mention results also concerning other types of algebraic structures.

An *algebra* will always mean a not necessarily associative algebra over a commutative ring with 1. Since rings are algebras over the ring of integers, we shall use the word "algebra" also for rings. For a subset B of an algebra A the ideal of A generated by B will be denoted by $\langle B\rangle_A$. For any algebra A we define

$$A^{(0)} = A \quad \text{and} \quad A^{(n)} = A^{(n-1)} \cdot A^{(n-1)}$$

for n = 1,2,... . A variety $\mathfrak{V}$ of algebras is called an *Andrunakievich variety of index* n if $\langle C\rangle_A^{(n)} \subseteq C$ for all $C \triangleleft B \triangleleft A$ and if n is the smallest such integer. If in the variety $\mathfrak{V}$ the relation $B \triangleleft A$ implies $B^s \triangleleft A$ for a fixed integer $s > 1$, then $\mathfrak{V}$ is said to be an s-variety. Important examples for Andrunakievich s-*varieties are*:

(1) The variety of associative algebras (index n = 2 and s = 2).

(2) An algebra A is said to be an *alternative* algebra, if $x^2y = x(xy)$ and $xy^2 = (xy)y$ for all $x,y \in A$. The variety of alternative algebras is an Andrunakievich 2-variety of index $n \leqq 15$. (Pchelintsev [48] proved that if $C \triangleleft B \triangleleft A$ then $(\langle C\rangle_A)^k \subseteq C$ where $k \leqq 4 \times 5^6$, and therefore $\langle C\rangle_A^{(15)} \subseteq C$.)

Further examples of Andrunakievich s-varieties are given in [2]; thus for instance, the 4-permutable algebras form an Andrunakievich 3-variety of index 1, the autodistributive algebras form an Andrunakievich 2-variety of index $n \leqq 4$.

A commutative R-algebra ($1/2 \in R$) A is called a *Jordan algebra* if $(x^2y)x = x^2(yx)$ holds for all $x,y \in A$. The variety of Jordan algebras is a 3-variety.

A (left) *near-ring* $A(+,\cdot)$ is a not necessarily commutative group $A(+)$, a semigroup $A(\cdot)$ and the left distributivity $x(y+z) = xy + xz$ $(x,y,z \in A)$ is satisfied. The near-ring A is called 0-symmetric if also $0.x = 0$ $(x \in A)$ holds. The variety of (0-symmetric) near-rings is neither an Andrunakievich nor an s-variety; nevertheless its radical theory has remarkable features.

An *involution* R-*algebra* A^* with involution * is an associative algebra subjected to the identities

$$(x+y)^* = x^* + y^*, \quad (xy)^* = y^*x^*$$

$$x^{**} = x, \qquad (\alpha x)^* = \alpha x^*$$

for all $x,y \in A$ and $\alpha \in R$. If R^* is also an involution R-algebra with involution * satisfying

$$(\alpha x)^* = \alpha^* x^*, \; x \in A, \; \alpha \in R,$$

instead of the last identity, then we speak of an *involution* R^*-*algebra*. The variety of involution R-algebras is an Andrunakievich 2-variety of index 2 (having one more operation, namely the involution, in this variety involution preserving homomorphisms and ideals closed under involution have to be considered).

Let $A \to B$ and $B \triangleleft A$ denote that B is a homomorphic image of A, and B is an ideal of A, respectively. Radical classes and semisimple classes in a variety of algebras can be defined in the following way:

A subclass $\subseteq \mathfrak{v}$ is called a *radical class* if it satisfies

(R) $A \in \rho \iff \forall\, A \to B \neq 0 \; \exists\, C \triangleleft B$ such that $0 \neq C \in \rho$.

A subclass $\sigma \subseteq \mathfrak{v}$ is called a semisimple class if it satisfies

(S) $A \in \sigma \iff \forall\, 0 \neq B \triangleleft A \; \exists\, B \to C$ such that $0 \neq C \in \sigma$.

Conditions (R) and (S) are dual to each other (by interchanging the relations $\to$ and $\triangleleft$). It is well-known that a subclass $\rho \subseteq \mathfrak{v}$ is a radical class if and only if:

(1) ρ is homomorphically closed;

(2) ρ is *inductive*: if $B_1 \subseteq \ldots \subseteq B_\lambda \subseteq \ldots$ is an ascending chain of ideals of an algebra A such that $B_\lambda \in \rho$ for each λ, then also $\cup B_\lambda \in \rho$;

(3) ρ is *closed under extensions*: $B \in \rho$ and $A/B \in \rho$ imply $A \in \rho$.

Let us notice that here for (1) one can demand the weaker condition

(1_0) if $A \in \rho$, then for every $A \to B \neq 0$ there exists a $C \triangleleft B$ such that $0 \neq C \in \rho$

(see [33], Theorem 4). Moreover, for a given radical class ρ, every algebra A has a largest ρ-ideal $\rho(A)$, called the *ρ-radical of* A. Every radical class determines the class

$$\mathcal{S}\rho = \{A \in \mathfrak{v}: \rho(A) = 0\}$$

and $\mathcal{S}\rho$ is the largest semisimple class σ such that $\sigma \cap \rho = 0$; further, every semisimple class σ determines the class

$$\mathcal{U}\sigma = \{A \in \mathfrak{v}: (A \to B \in \sigma \Rightarrow B = 0)\}$$

which is the largest radical class ρ such that $\rho \cap \sigma = 0$.

2. THE ADS PROPERTY

Though the defining conditions (R) and (S) of radical and semisimple classes are duals of each other, we cannot expect, in general, dual theorems for radical and semisimple classes, unless the variety $\mathfrak{v}$ possesses certain additional properties. The reason for that state of affairs is that the relation $\to$ is transitive, but the relation $\triangleleft$ is not. A remedy for the missing transitivity of the relation $\triangleleft$ is the so-called ADS property (where ADS stands for Anderson-Divinsky-Suliński) of a radical class $\rho \subseteq \mathfrak{v}$:

(ADS) $\forall\, B \triangleleft A \in \mathfrak{v} \Rightarrow \rho(B) \triangleleft A$.

Recently Beidar [14] has proved the following result.

<u>THEOREM 1</u>: If the variety $\mathfrak{V}$ satisfies conditions:

(B1) to every $C \triangleleft B \triangleleft A \in \mathfrak{V}$ and $a \in A$ there exists an integer $m \geq 1$ such that $aC^{(m)} + C^{(m)}a \subseteq C$;

(B2) to every $c \triangleleft B \triangleleft A \in \mathfrak{V}$ and $a \in A$ there exists an integer $n \geq 1$ such that $(\langle aC + Ca + C\rangle_B)^{(n)} \subseteq C$;

then every radical class $\rho \subseteq \mathfrak{V}$ has the ADS property.

Beidar [14] has shown that the varieties of associative and alternative algebras, of Jordan R-algebras ($1/2 \in R$), of (γ,δ)-algebras over R ($1/6 \in R$) and the Andrunakievich s-varieties satisfy conditions (B1) and (B2), recovering the ADS theorem for associative and alternative algebras [1], and sharpening results of Nikitin [46], Markovichev [42], Anderson and Gardner [2], and Añh, Loi and Wiegandt [8].

The conclusion of Theorem 1 can be obtained also by imposing another, more category-theoretical, assumption. This is due to Terlikowska-Osłowska [60], [61]. Here we give a version of Terlikowska-Osłowska's condition as modified by Puczyłowski [50] and Veldsman [65].

An ideal B of an algebra $A \in \mathfrak{V}$ is called a *distinguished ideal* (denoted by $B \blacktriangleleft A$) if A does not contain ideals C and D such that $C \subsetneqq B \subsetneqq D$, $B/C \cong D/B$ and in the algebra B/C the relation $\triangleleft$ is transitive. By [61] Corollary 1 (cf. also [49] and [65]) we have

<u>THEOREM 2</u>: If a variety $\mathfrak{V}$ of algebras satisfies condition

(T0) the variety $\mathfrak{V}$ is *normal* : $C \blacktriangleleft B \triangleleft A \in \mathfrak{V} \Rightarrow C \triangleleft A$, then every radical class $\rho \subseteq \mathfrak{V}$ has the ADS property.

Terlikowska-Osłowska [60] proved that the varieties of associative and alternative algebras satisfy condition (T0), recovering the ADS theorem [1]. Further examples are the varieties of Γ-rings, Nobusawa's cubic rings and autodistributive rings, as has been shown by Veldsman [65]. Puczyłowski

[50] has shown that semigroups with 0 (considering Rees factor semigroups for the relation $\rightarrow$) satisfy condition (TO). The variety of involution algebras is, of course, not a normal variety, as in view of [38] and [39] not every radical class of involution algebras has the ADS property. Nevertheless, in the definition of distinguished ideals of involution algebras Puczyłowski [50] used the relation A^* is isomorphic to B^* or $A^2 = B^2 = 0$ and A^* is isomorphic to B^{-*} (where $b^{-*} = -b^*$, $b \in B$) instead of the relation $\cong$, and he has shown that involution R^*-algebras satisfy condition (TO). Thus he has got the sufficiency part of Loi's characterization [38]: a radical class ρ of involution R^*-algebras has the ADS property if and only if

(L) $\quad A^* \in \rho$ and $A^2 = 0 \Rightarrow A^{-*} \in \rho$.

In view of this result we conclude that Beidar's conditions (B1) and (B2) cannot be directly applied to involution R^*-algebras (because not every radical has the ADS property), though involution R^*-algebras form an Andrunakievich 2-variety (which is, however, not a subvariety of all not necessarily associative rings).

In [27], Proposition 1.2, Krempa proved that in the variety of all Lie algebras a radical ρ has the ADS property if and only if for every Lie algebra A its radical $\rho(A)$ is a characteristic ideal (that is $d(\rho(A)) \subseteq \rho(A)$ for every derivation d on A).

So far many attempts have been made to carry over an ADS property theorem for near-rings without success. Nevertheless, the Brown-McCoy radical and the J_2 and J_3 radicals - generalizations of the Jacobson radical for near-rings - do have the ADS property (this follows immediately from results of [3] and [45]).

In the variety of rings graded by an abelian group every radical class has the ADS property [59], but for rings graded by a group or semigroup this is not necessarily so (cf. Krempa and Terlikowska-Osłowska [28]).

Let us mention that the ADS property of a radical class ρ is equivalent to the following condition imposed on its semisimple class σ [45]:

if $C \triangleleft B \triangleleft A \in \mathfrak{V}$ and $B/C \in \sigma$, then there exists a $D \triangleleft A$ such that $D \subseteq C$ and $B/D \in \sigma$.

3. THE HEREDITARINESS OF SEMISIMPLE CLASSES

An immediate consequence of the ADS property of a radical class ρ is that the corresponding semisimple class σ is hereditary: $B \lhd A \in \sigma \Rightarrow \in \sigma$. Thus by Beidar [14] in an Andrunakievich s-variety every semisimple class is hereditary (cf. also [2] and [8]). For Jordan R-algebras ($1/2 \in R$) an analogous result was proved by Nikitin [46] and for (γ,δ)-algebras over ($1/6 \in R$) by Markovichev [42].

In the variety $\mathfrak{V}$ of all not necessarily associative algebras a semisimple class of a radical class is hereditary if and only if the radical property depends only on the structure of the underlying group of the algebra. This degenerating result has been discovered by Gardner [18]: a semisimple class $\sigma \subseteq \mathfrak{V}$ is hereditary if and only if the radical class $\rho = \mathcal{U}\sigma$ satisfies the following condition:

$$A \in \rho \quad \text{and} \quad A(+) \cong B(+) \Rightarrow B \in \rho$$

(cf. also [30], [31]). In [18] Gardner investigated also the case of commutative, anticommutative and power associative algebras.

In the variety $\mathfrak{V}$ of near-rings the hereditariness of a semisimple class σ forces that the corresponding radical class $\rho = \mathcal{U}\sigma$ contains all near-rings A with $A^2 = 0$. This result was proved by Betsch and Kaarli [15] (see also [16]). The converse, however, is not true: the semisimple class of the nil-radical of near-rings is not hereditary [23], [63]. As in the case of associative algebras, several important radical classes ρ of near-rings are *left strong* (that is, the radical $\rho(A)$ contains every ρ-radical left ideal of A); for instance certain generalizations of the Jacobson radical for near-rings (see [4]). Mlitz and Oswald [44] proved that the semisimple classes of any left strong radical of near-rings is hereditary, provided that the semisimple class contains a 0-symmetric near-ring $\neq 0$.

One may ask whether the hereditariness of a semisimple class σ implies that the radical class $\rho = \mathcal{U}\sigma$ has the ADS property. As Barry Gardner remarked (private communication) from Krempa [27] one can get an example for a class of finite-dimensional Lie algebras in which the hereditariness of a semisimple class does not imply the ADS property of its radical class.

Let us mention that the hereditariness of a semisimple class σ is equivalent to the following condition on its radical class ρ (see [29] and

[11]):

if $C \triangleleft B \triangleleft A \in \mathfrak{v}$ and $C \in \rho$, then $\langle C\rangle_A \in \rho$.

4. CHARACTERIZATIONS OF SEMISIMPLE CLASSES

The lack of transitivity of the relation $\triangleleft$ was for long an obstacle in characterizing semisimple classes by transparent algebraic conditions dual to those of (1), (2) and (3) in the characterization of radical classes. Several characterizations of semisimple classes of algebras are known (cf. for instance, [34], [35], [43]); here we shall recall two of them. First of all, let us mention that the dual condition of (1), i.e. the hereditariness of semisimple classes, is not a property shared by all semisimple classes (see [30]). Instead, we have the necessity part of condition (S) which is exactly the dual condition to (1_0).

For a subclass σ of a variety of algebras the following conditions are equivalent:

(I) σ is semisimple class;

(II)(a) σ is *regular*: to every $A \in \sigma$ and $A \to B \neq 0$ there exists $C \triangleleft B$ such that $0 \neq C \in \sigma$,

(b) σ is closed under subdirect sums,

(c) σ is closed under extensions,

(d) $((A)\sigma)\sigma \triangleleft A$ for every $A \in \mathfrak{v}$ where $(X)\sigma$ is defined as $(X)\sigma = \cap(B \triangleleft X : X/B \in \sigma)$;

(III)(a) σ is regular,

(b_0) σ is *coinductive* : if $B_1 \supseteqq \dots \supseteqq B_\lambda \supseteqq \dots$ is a descending chain of ideals of $A \in \mathfrak{v}$ such that each A/B_λ is in σ, then also $A/\cap B_\lambda \in \sigma$,

(c) σ is closed under extensions,

(d) if $C \triangleleft B \triangleleft A \in \mathfrak{v}$ and B and C are minimal with respect to $A/B \in \sigma$ and $B/C \in \sigma$, respectively, then $C \triangleleft A$.

In these characterizations condition (d) seems to be more elegant, but in fact it is easier to deal with condition (e) (cf. [6], [9], [41], [54], [55], [65]). In order to establish a complete duality between radical and semisimple classes one has to eliminate condition (e) in characterization (III). For associative algebras this was done by Sands [55]. Sands' theorem has been extended to alternative rings in [6]. Recently Beidar [14] has proved the following generalization of Sands' theorem:

THEOREM 3: Let $\mathfrak{V}$ be a variety of algebras satisfying conditions (B1) and (B2) and

(B3) to every $C \lhd B \lhd A \in \mathfrak{V}$ there exists an integer $t \geq 1$ such that $\langle C^{(t)} \rangle_B \subseteq C^2$.

A subclass $\sigma \subseteq \mathfrak{V}$ is a semisimple class if and only if σ is regular, coinductive and closed under extensions.

He has shown in [14] that associative algebras, alternative algebras, Jordan R-algebras ($1/2 \in R$), (γ,δ)-algebras ($1/6 \in R$), and Andrunakievich s-varieties also satisfy condition (B3), sharpening results of [2] and [8].

The following counterpart of Theorem 3 has been proved by Puczyłowski [50] and Veldsman [65] (see also [49]).

THEOREM 4: Let $\mathfrak{V}$ be a variety of algebras which is normal (that is, satisfies condition (T0) of Theorem 1). A subclass $\sigma \subseteq \mathfrak{V}$ is a semisimple class if and only if σ is regular, coinductive and closed under extensions.

Thus, with the necessary modification in the definition of distinguished ideals [50], Theorem 4 recovers also Loi's characterization of those semisimple classes of involution algebras whose radical class satisfies condition (L) (cf. [36] and [37]).

Sands' theorem characterizing semisimple classes has been extended to Hausdorff topological rings [41]. For characterization of semisimple classes of compact rings we refer to [7].

In Theorems 3 and 4 regularity can be replaced by hereditariness of σ (cf. [8], [9], [33], [54], [59], [61]).

5. SEMISIMPLE CLASSES OF RADICALS SATISFYING ADDITIONAL REQUIREMENTS

The classical radicals usually possess certain extra properties, such as being hereditary, supernilpotent, special, etc. Theorems 3 and 4 enable us to characterize the semisimple classes of such radicals. In this section, therefore, we assume that the variety considered satisfies conditions (B1), (B2), (B3) or (T0). One can derive the following generalization of van Leeuwen's theorem [32] from Theorems 3 and 4.

THEOREM 5: A subclass σ of the variety $\mathfrak{v}$ is the semisimple class of a hereditary radical class ρ if and only if σ is regular, closed under subdirect sums and essential extensions, that is, if B is an essential ideal of A and $B \in \sigma$, then also $A \in \sigma$ (B is an essential ideal of A, if $B \cap C \neq 0$ for every $0 \neq C \triangleleft A$).

A radical class $\rho \subseteq \mathfrak{v}$ is called *supernilpotent* if ρ is hereditary and contains all algebras A with $A^2 = 0$. Assuming that the variety $\mathfrak{v}$ contains all algebras A with $A^2 = 0$, from Theorem 5 one can get the following result (see [5], [9]).

THEOREM 6: A subclass σ of the variety $\mathfrak{v}$ is the semisimple class of a supernilpotent radical class ρ if and only if σ is regular, closed under subdirect sums and essential extensions and is *weakly homomorphically closed*: if $B \triangleleft A \in \mathfrak{v}$ and $B^2 = 0$, then $A/B \in \sigma$.

Here the condition "σ is weakly homomorphically closed" can be replaced by demanding "σ consists of semiprime algebras" (see [53]). An analogous result is valid also for Hausdorff topological rings (see [41]). For near-rings Veldsman [62] has proved that a radical class is *hypersolvable* (that is, $A \in \mathfrak{v}$ and $A^2 = 0$ imply $A \in \sigma$) if and only if the corresponding semisimple class is weakly homomorphically closed.

A radical class ρ is *subidempotent* if ρ is hereditary and consists of idempotent algebras. Veldsman [64] has proved the following:

THEOREM 7: A subclass σ of the variety $\mathfrak{v}$ is the semisimple class of a subidempotent radical class ρ if and only if σ is regular, closed under subdirect sums and essential extensions and satisfies the following condition if $C \triangleleft B \triangleleft A \in \mathfrak{V}$ and $B = (B)\sigma$, then $C \triangleleft A$.

Considering a variety $\mathfrak{V}$ of K-algebras over a field K every radical class ρ is either hypersolvable (that is, ρ contains all algebras $A \in \mathfrak{V}$ with $A^2 = 0$) or *hypoidempotent* (that is, ρ consists of idempotent algebras), as has been shown in [2]. The same assertion is also valid surprisingly in the class of all Hausdorff topological rings (with continuous homomorphisms), as has been proved long ago by Arnautov and Vodinchar [12].

Veldsman [64] has proved that in any variety of algebras a radical class is hypersolvable or hypoidempotent if and only if its semisimple class σ satisfies the following condition:

if $B \triangleleft A \in \sigma$ and $A^2 = 0$, then $A/B \in \sigma$,

and has also obtained characterizations of semisimple classes of hypo-idempotent radicals.

Those radicals which provide a subdirect decomposition theorem for semisimple algebras are the special radicals of Andrunakievich [10]. A subclass $\mu \subseteq \mathfrak{V}$ is said to be *special class* if μ is a hereditary class closed under essential extensions and consisting of prime algebras (cf. also [21]). The radical class $\rho = \mathcal{U}\mu$ is called a *special radical*. An algebra $A \in \mathfrak{V}$ is semisimple with respect to the special radical ρ if and only if A is a subdirect sum of prime algebras each in $\sigma = \mathcal{S}\rho$. By [19] a subclass ρ of associative or alternative algebras is a special radical class if and only if ρ is homomorphically closed, hereditary and satisfies:

if every nonzero factor algebra B of $A \in \mathfrak{V}$ which is a prime algebra has an ideal C such that $0 \neq C \in \rho$, then $A \in \rho$.

The semisimple class of a special radical has been characterized as a subclass σ being regular, closed under subdirect sums and essential extensions and satisfying:

if $A \in \sigma$, then A is a subdirect sum of prime algebras each in σ (see [53]).

In carrying over results on special radicals to varieties of nonassociati algebras and near-rings we mention the papers [3], [17], [22], [24], [45]. Let τ be a regular and essentially closed subclass of a variety $\mathfrak{v}$ of not necessarily associative algebras. If τ satisfies the condition:

(F) $\quad C \lhd B \lhd A \in \mathfrak{v}$ and $B/C \in \tau \Rightarrow C \lhd A$,

then every algebra A in the semisimple class $\sigma = \mathcal{SU}\tau$ is a subdirect sum of algebras each in τ. In the variety $\mathfrak{v}$ and also in the variety of near-rings any regular class satisfying condition (F) consists of semiprime algebras (or near-rings, respectively). In an Andrunakievich variety the class of all semiprime algebras satisfies condition (F); this is not true for near-rings (cf. [23]). A special feature of the variety of near-rings (as well as of associative and alternative algebras) is that it satisfies condition

(G) $\quad$ if $C \lhd B \lhd A$ and $1 \in B/C$ then $C \lhd A$

(see [3]). Condition (G) makes it possible to define Brown-McCoy type radicals for near-rings having all the specific properties of special radicals. Nevertheless, as has been shown recently by Kaarli and Kriis [25], the Baer radical of near-rings (that is, the upper radical of all prime near-rings) is not a Kurosh-Amitsur radical.

6. THE LOWER RADICAL CONSTRUCTION

In a variety $\mathfrak{v}$ the smallest radical class containing a given subclass can be constructed as follows. Let be a homomorphically closed subclass of $\mathfrak{v}$, and define

$$\gamma_1 = \gamma$$

$$\gamma_{\lambda+1} = \{A \in \mathfrak{v}: A \to B \neq 0 \Rightarrow \exists\, C \lhd B \text{ such that } 0 \neq C \in \gamma_\lambda\},$$

and

$$\gamma_\lambda = \cup(\gamma_\nu: \nu < \lambda) \text{ for limit ordinals } \lambda.$$

Then

$$L\gamma = \cup(\gamma_\lambda : \text{all ordinals } \lambda)$$

is the smallest radical class containing the subclass γ. Concerning this lower radical construction, which goes back to Kurosh, two interesting questions can be asked:

(1) Given a variety $\mathfrak{V}$ of algebras, does there exist an ordinal ζ such that the lower radical construction terminates in at most ζ steps, that is $L\gamma = \cup(\gamma_\lambda : \lambda \leqq \zeta)$?

(2) Given a variety $\mathfrak{V}$ of algebras and an ordinal n, does there exist a homomorphically closed subclass $\gamma \subseteq \mathfrak{V}$ such that the lower radical construction terminates in exactly n steps, that is $L\gamma = \gamma_n$ but $L\gamma \neq \gamma_k$ for $k < n$?

In answering these problems, many interesting properties of algebras and of their ideal structure have been discovered. We shall mention some of them here:

(A) Let $\mathfrak{V}$ be the variety of all (not necessarily associative) R-algebras, or that of all commutative R-algebras, or that of all anticommutative R-algebras. If $\gamma = \gamma_1$ is a homomorphically closed subclass of $\mathfrak{V}$, which is not a radical class, then $L\gamma \neq \gamma_\lambda$ for every ordinal λ ([14], cf. also [11] [51] and [52]).

The following general theorem is due to Beidar [14] and Puczyłowski [49] (and in a slightly weaker form to Veldsman [64]).

<u>THEOREM 8</u>: Let $\mathfrak{V}$ be a variety of algebras and assume that either $\mathfrak{V}$ satisfies conditions (B1), (B2) and in (B2) the integer n depends only on Λ, or $\mathfrak{V}$ satisfies condition (T0). Then for every homomorphically closed class the lower radical construction terminates at the first limit ordinal ω_0, that is $L\gamma = \gamma_{\omega_0}$.

This theorem was first proved for associative rings by Suliński, Anderson and Divinsky [58], and Heinicke [20] showed that ω_0 is the lowest upper bound. The result of [58] has been extended for alternative rings by Krempa [26], for Jordan algebras and (γ,δ)-algebras by Nikitin [47].

In proving his result Krempa [26] used the following interesting lemma:

LEMMA 1: If $C \triangleleft B \triangleleft A$ and $B = \langle C \rangle_A$, then every nonzero homomorphic image of B contains a nonzero ideal which is a homomorphic image of C. (Here the algebras are alternative.)

(B) Question (2), which is called the Suliński-Anderson-Divinsky problem, was open for 16 years, and has been solved by Beidar [13]. Here we present the more general result due to L'vov and Sidorov [40] (cf. also [14]).

THEOREM 9: Let $\mathfrak{V}$ be a variety of associative R-algebras containing the variety of all commutative R-algebras. To every natural number n = 1,2,... and to $n = \omega_0$ there exists a homomorphically closed class $\gamma = \gamma_1$ such that the lower radical construction terminates in exactly n steps.

In proving this result for associative rings and in giving a positive answer to the Suliński-Anderson-Divinsky problem, Beidar [13] made use of interesting properties of the Gaussian integers G.

LEMMA 2: Let p be a prime and

$$A_n = \{pa + p^n b : a,b \text{ integers}\}$$

for n = 1,2,... . If R is a subring of G and $A_{n+1} \triangleleft R$, then either $R = A_{n+1}$ or $R = A_n$ or $1 \in R$.

Interesting properties of certain algebras and semigroups were discovered by L'vov and Sidorov [40] and by Sidorov [56], respectively. An interesting feature of the ideal structure of associative algebras, which is useful in proving results concerning the terminating of lower radical construction, is due to Stewart [57].

$$0 \neq B = B_1 \triangleleft \dots \triangleleft B_n \quad A \quad (n \geq 1)$$

and $B \in \delta$, then there exists an algebra $C \in \delta$ such that $C^2 = 0$ and $0 \neq C \triangleleft D \triangleleft A$.

In proving characterizations of semisimple classes and that the Kurosh lower radical construction terminates at most at the first limit ordinal, conditions (B1), (B2) and (B3) and condition (T0) played a decisive role. Thus the following question arises: Does there exist any connection between conditions (B1), (B2) and (B3) and condition (T0)?

ACKNOWLEDGEMENTS

The author gratefully acknowledges the financial support of the Australian Mathematical Society and of the Hungarian National Foundation for Scientific Research Grant No. 1813.

REFERENCES

[1] T. Anderson, N. Divinsky and A. Suliński, Hereditary radicals in associative and alternative rings, Canad. J. Math. 17 (1965), 594-603.

[2] T. Anderson and B.J. Gardner, Semi-simple classes in a variety satisfying an Andrunakievich Lemma, Bull. Austral. Math. Soc. 18 (1978), 187-200.

[3] T. Anderson, K. Kaarli and R. Wiegandt, Radicals and subdirect decompositions, Commun. Algebra 13 (1985), 479-494.

[4] T. Anderson, K. Kaarli and R. Wiegandt, On left strong radicals of near-rings, Proc. Edinb. Math. Soc. 31 (1988), 447-456.

[5] T. Anderson and R. Wiegandt, Weakly homomorphically closed semisimple classes, Acta Math. Acad. Sci. Hung. 34 (1979), 329-336.

[6] T. Anderson and R. Wiegandt, Semisimple classes of alternative rings, Proc. Edinb. Math. Soc. 25 (1982), 21-26.

[7] P.N. Ánh, On semisimple classes of topological rings, Ann. Univ. Sci. Budapest 20 (1977), 59-70.

[8] P.N. Ảnh, N.V. Loi and R. Wiegandt, On the radical theory of Andrunakievich varieties, Bull. Austral. Math. Soc. 31 (1985), 257-269.

[9] P.N. Ánh and R. Wiegandt, Semisimple classes of non-associative rings and Jordan algebras, Commun. Algebra 13 (1985), 2669-2690.

[10] V.A. Andrunakievich, Radicals of associative rings (Russian) I, Mat. Sb. 44 (1958), 179-212 and II, Mat. Sb. 55 (1961), 329-346.

[11] V.A. Andrunakievich and Ju. M. Rjabuhin, Torsions and Kurosh chains in algebras (Russian), Trudy Mosk. Mat. Obshch. 29 (1973), 19-49.

[12] V.I. Arnautov and M.I. Vodinchar, Radicals of topological rings (Russian), Mat. Issled. Kishinev 3, No. 2 (1968), 31-61.

[13] K.I. Beidar, A chain of Kurosh may have an arbitrary finite length, Czech. Math. J. 32 (1982), 418-422.

[14] K.I. Beidar, Semisimple classes of algebras and the lower radical (Russian), Mat. Issled. Kishinev to appear.

[15] G. Betsch and K. Kaarli, Supernilpotent radicals and hereditariness of semisimple classes of near-rings, Colloq. Math. Soc. J. Bolyai, 38, Radical Theory, Eger, 1982, North-Holland, Amsterdam, 1985, pp. 47-58.

[16] G. Betsch and R. Wiegandt, Non-hereditary semisimple classes of near-rings, Stud. Sci. Math. Hung. 17 (1982), 69-75.

[17] A. Buys and G. Gerber, Special classes in Ω-groups, Ann. Univ. Sci. Budapest 29 (1986), 73-85.

[18] B.J. Gardner, Some degeneracy and pathology in non-associative radical theory I, Ann. Univ. Sci. Budapest 22/23 (1979/80), 65-74 and II, Bull Austral. Math. Soc. 23 (1981), 423-428.

[19] B.J. Gardner and R. Wiegandt, Characterizing and constructing special radicals, Acta Math. Acad. Sci. Hung. 40 (1982), 73-83.

[20] A. Heinicke, A note on lower radical construction, Canad. Math. Bull. 11 (1968), 23-30.

[21] G.A.P. Heyman and C. Roos, Essential extensions in radical theory for rings, J. Austral. Math. Soc. 23 (1977), 340-347.

[22] M. Holcombe and R. Walker, Radicals in categories, Proc. Edinb. Math. Soc. 21 (1978), 111-128.

[23] K. Kaarli, Classification of irreducible R-groups over semiprimary near-rings (Russian), Tartu Riikl. Ül. Toimetised 556 (1981), 47-63.

[24] K. Kaarli, Special radicals of near-rings (Russian), Tartu Riikl. Ül. Toimetised 610 (1982), 53-68.

[25] K. Kaarli and T. Kriis, Prime radical of near-rings, Tartu Riikl. Ül. Toimetised 764 (1987), 23-29.

[26] J. Krempa, Lower radical properties for alternative rings, Bull. Acad. Polon. Sci. 23 (1975), 139-142.

[27] J. Krempa, Radicals and derivation algebras, Colloq. Math. Soc. J. Bolyai, 38, Radical Theory, Eger, 1982, North-Holland, Amsterdam, 1985, pp. 195-227.

[28] J. Krempa and B. Terlikowska-Osłowska, On graded radical theory, Preprint.

[29] W.G. Leavitt, Strongly hereditary radicals, Proc. Amer. Math. Soc. 21 (1969), 703-705.

[30] W.G. Leavitt and E.P. Armendariz, Nonhereditary semi-simple classes, Proc. Amer. Math. Soc. 18 (1967), 1114-1117.

[31] W.G. Leavitt and R. Wiegandt, Torsion theory for not necessarily associative rings, Rocky Mountain J. Math. 9 (1979), 259-271.

[32] L.C.A. van Leeuwen, Properties of semisimple classes, J. Nat. Sci. Math., Lahore 15 (1975), 59-67.

[33] L.C.A. van Leeuwen, C. Ross and R. Wiegandt, Characterizations of semisimple classes, J. Austral. Math. Soc. 23 (1977), 172-182.

[34] L.C.A. van Leeuwen and R. Wiegandt, Radicals, semi-simple classes and torsion theories, Acta Math. Acad. Sci. Hung. 36 (1980), 37-47.

[35] L.C.A. van Leeuwen and R. Wiegandt, Semisimple and torsionfree classes, Acta Math. Acad. Sci. Hung. 38 (1981), 73-81.

[36] N.V. Loi, On the radical theory of involution algebras, Acta Univ. Carol. 27 (1986), 29-40.

[37] N.V. Loi, A note on the radical theory of involution algebras, Stud. Sci. Math. Hung. 23 (1988), 157-160.

[38] N.V. Loi, The A-D-S property for radicals of involution K-algebras, Arch. Math. 49 (1987), 196-199.

[39] N.V. Loi and R. Wiegandt, Involution algebras and the Anderson-Divinsky-Suliński property, Acta Sci. Math. Szeged. 50 (1986), 5-14.

[40] I.V. L'vov and A.V. Sidorov, On the stabilization of Kurosh chains (Russian), Mat. Zametki 36 (1984), 815-821.

[41] L. Márki, R. Mlitz and R. Wiegandt, A note on radical and semisimple classes of topological rings, Acta Sci. Math. Szeged. 51 (1987), 145-151.

[42] A.S. Markovichev, On the hereditariness of radicals of rings of type (γ,δ) (Russian), Alg. Logika 17 (1978), 33-55.

[43] R. Mlitz, Radicals and semisimple classes of Ω-groups, Proc. Edinb. Math. Soc. 23 (1980), 37-41.

[44] R. Mlitz and A. Oswald, Hypersolvable and supernilpotent radicals of near-rings, Stud. Sci. Math. Hung. 24 (1989), to appear.

[45] R. Mlitz and S. Veldsman, Radicals and subdirect decomposition of Ω-groups, Preprint.

[46] A.A. Nikitin, On the hereditariness of radicals of rings (Russian), Alg. Logika 17 (1978), 303-315.

[47] A.A. Nikitin, On lower radicals of rings (Russian), Alg. Logika 17 (1978), 596-610.

[48] S.V. Pchelintsev, On metaideals of alternative algebras (Russian), Sib. Mat. Zh. 24 (1983), 142-148.

[49] E.R. Puczyłowski, On semisimple classes of associative and alternative rings, Proc. Edinb. Math. Soc. 27 (1984), 1-5.

[50] E.R. Puczyłowski, On general theory of radicals, Preprint.

[51] Ju. M. Rjabuhin, On lower radicals of rings (Russian), Mat. Zametki 2 (1967), 239-244.

[52] Ju. M. Rjabuhin, The theory of radicals in non-associative rings (Russian), Mat. Issled. Kishinev 3, No. 7 (1968), 86-99.

[53] Ju. M. Rjabuhin and R. Wiegandt, On special radicals, supernilpotent radicals and weakly homomorphically closed classes, J. Austral. Math. Soc. 31 (1981), 152-162.

[54] A.D. Sands, Strong upper radicals, Q. J. Math. Oxford 27 (1976), 21-24.

[55] A.D. Sands, A characterisation of semisimple classes, Proc. Edinb. Math. Soc. 24 (1981), 5-7.

[56] A.V. Sidorov, On stabilization of Kurosh chains in the class of semigroups with 0 (Russian), Sib. Mat. Zh. 29 (1988), 131-136.

[57] P.N. Stewart, On the lower radical construction, Acta Math. Acad. Sci. Hung. 25 (1974), 31-32.

[58] A. Suliński, T. Anderson and N. Divinsky, Lower radical properties for associative and alternative rings, J. Lond. Math. Soc. 41 (1966), 417-424.

[59] A. Suliński and R. Wiegandt, Radicals of rings graded by abelian groups, Colloq. Math. Soc. J. Bolyai, 38, Radical Theory, Eger, 1982, North-Holland, Amsterdam, 1985, pp. 607-610.

[60] B. Terlikowska-Osłowska, Category with a self-dual set of axioms, Bull. Acad. Polon. Sci. 25 (1977), 1207-1214.

[61] B. Terlikowska-Osłowska, Radical and semisimple classes of objects in categories with a self-dual set of axioms, Bull. Acad. Polon. Sci. 26 (1978), 7-13.

[62] S. Veldsman, Supernilpotent radicals of near-rings, Commun. Algebra 15 (1987), 2497-2509.

[63] S. Veldsman, On the non-hereditariness of radical and semisimple classes of near-rings, Preprint.

[64] S. Veldsman, Subidempotent radical classes, Quaest. Math. 11 (1988), 361-370.

[65] S. Veldsman, Sufficient conditions for a well-behaved Kurosh-Amitsur radical theory, Proc. Edinb. Math. Soc. to appear.

R. Wiegandt
Mathematical Institute of the
Hungarian Academy of Sciences
P.O. Box 127
H-1364 Budapest
Hungary

ZHU YUAN-SEN

Brown-McCoy semisimplicity of group rings

Let R(G) denote a group ring of a group G over a ring R. Semisimplicity of group rings is one of the central problems in studying group rings. Some good results on the semisimplicity of group algebras have been obtained [2]. Some of the results have been extended to commutative rings in [1], and also generalized to general associative rings in [3]. However, the problem has not been solved yet for any group. Brown-McCoy semisimplicity of finitely generated torsion-free abelian groups over a commutative ring and that of locally finite ZA-groups and residually ZA-groups over a field have been discussed in [8] and [9] respectively. On the other hand, for some special groups, it has been proved that the J-radical of F(G) over a field F may be controlled by some subgroups of G in [4] and [5], namely JF(G) = JF(H)F(G). Further, H.K. Farahat put forward the question when does the equality JR(G) = (JR)(G) hold? So the J-radical problem of R(G) becomes that of R. It has been proved in [6] and [7] that Farahat's equality is true for the J-radical when G is locally finite and R is either a semiprimary ring or a commutative ring.

Mainly Brown-McCoy semisimplicity of group rings over an associative ring is studied in this paper. We have solved the Brown-McCoy semisimplicity problem of R(G), when G is an abelian group, a finite group or a residually finite group. Brown-McCoy semisimple, in its abbreviated form, is *G*-semisimple. It has been proved that Farahat's equality holds for the *G*-radical if G is a finite group or Abelian torsion group, R is associative or commutative ring respectively.

1. ABELIAN GROUPS

We all know that the *G*-radical is equal to the J-radical when R is a commutative ring. By [1], Theorem 6 and 7, the following theorem can be obtained immediately.

THEOREM 1: Suppose that R is a commutative ring with 1.

(i) If G is an abelian torsion group, then $R(G)$ is G-semisimple if and only if R is G-semisimple, and each $n \in O(G)$ ($O(G)$ is the set of orders of subgroups of the group G) is regular in R.

(ii) If G is a torsion-free abelian group, then $R(G)$ is G-semisimple if and only if R is semiprime and each $n \in O(G)$ is regular in R.

(iii) If G is a locally finite group and R is G-semisimple and each $n \in O(G)$ is regular in R, then $R(G)$ is J-semisimple.

If both R and G are commutative, then $GR(G) = JR(G)$; therefore we only discuss the case that there is at least one noncommutative in both R and G.

Suppose that R is an associative ring with 1 and H is a subgroup of G. Let Π_H be a transversal map of $R(G)$ into $R(H)$ given by

$$\Pi_H(\alpha) = \Pi_H(\sum_{g\in G} a_g g) = \sum_{g\in H} a_g g.$$

It is easy to show that

$$\Pi_H(a\alpha + b\beta) = a\Pi_H(\alpha) + b\Pi_H(\beta),$$

$$\Pi_H(\nu\alpha) = \nu\Pi_H(\alpha), \qquad \Pi_H(\alpha\nu) = \Pi_H(\alpha)\nu,$$

where

$$\alpha,\beta \in R(G), \quad \nu \in R(H), \quad a,b \in R.$$

LEMMA 1.1: Suppose that R is an associative ring with 1 and H is a subgroup of a group G. If H is contained in the centre of G, then $R(H) \cap GR(G) \subset GR(H)$. Particularly, $R \cap GR(G) \subset GR$.

PROOF: Let Y be a representative set of right cosets of H in G. Then every element $\beta \in R(G)$ can be expressed uniquely in the form of a finite sum:

$$\beta = \sum_{g\in Y} \beta_g\, g, \quad \text{where} \quad \beta_g \in R(H), \quad g \in Y.$$

Since $R(H) \cap GR(G) \triangleleft R(H)$, for any $\alpha \in R(H) \cap GR(G)$, thus $\alpha \in G_{R(G)}(\alpha)$

and there exist $x, x_i, y_i \in R(G)$ such that

$$\alpha = x + \alpha x + \sum_i (x_i \alpha y_i + x_i y_i)$$

where

$$x_i = \sum \alpha_{ij} g_j, \ y_i = \sum \beta_{ij} g_j, \quad \alpha_{ij}, \beta_{ij} \in R(H).$$

Then

$$\alpha = x + \alpha x + \sum(\alpha_{ij} g_j \alpha \beta_{i\ell} g_\ell + \alpha_{i\ell} g_j \beta_{i\ell} g_\ell)$$

$$= x + \alpha x + \sum(\alpha_{ij} \alpha \beta_{i\ell} g_j g_\ell + \alpha_{ij} \beta_{i\ell} g_j g_\ell).$$

Let $\Pi_H : R(G) \to R(H)$ be a transversal map; hence

$$\alpha = \Pi_H(x) + \alpha \Pi_H(x) + \sum[\Pi_H(\alpha_{ij} \alpha \beta_{i\ell} g_j g_\ell) + \Pi_H(\alpha_{ij} \beta_{i\ell} g_j g_\ell)]$$

$$= \Pi_H(x) + \alpha \Pi_H(x) + \sum(\alpha_{ij} \alpha \beta_{ij'} + \alpha_{ij} \beta_{ij'}),$$

where j' is the correspondence subscript when $g_e = g_j^{-1}$. Whence there is $\alpha \in G_{R(H)}(\alpha)$ and α is also G-regular in $R(H)$, that is, $\alpha \in GR(H)$. Therefore, $R(H) \cap GR(G) \subset GR(H)$. □

THEOREM 1.2: Suppose that R is an associative ring with 1 and G is a torsion-free abelian group. If R is G-semisimple, then R(G) is also G-semisimple.

PROOF: Let $H = \langle \text{Supp } \alpha \rangle$ for any $\alpha \in GR(G)$. Clearly H is a finitely generated torsion-free abelian group. By the theorem in [8], we know that $GR(H) = 0$. However $\alpha \in R(H) \cap GR(G) \subset GR(H)$ by Lemma 1.1; then $GR(G) = 0$. □

LEMMA 1.2: If R is a subdirect sum of rings R_α $(\alpha \in \Omega)$ and G is any group, then R(G) is a subdirect sum of $R_\alpha(G)$ $(\alpha \in \Omega)$, which is denoted by

$$R(G) \sum_{\alpha \in \Omega}{}_S \oplus R_\alpha(G).$$

PROOF: Since $R = \Sigma_s \oplus R_\alpha$, let $\gamma_\alpha: R \to R_\alpha$ be projective and $I_\alpha = \text{Ker } \gamma_\alpha$, then $\bigcap_{\alpha\in\Omega} I_\alpha = 0$ and $R_\alpha \simeq R/I_\alpha$.

We extend γ_α to $\tilde{\gamma}_\alpha: R(G) \to R_\alpha(G)$ such that $\tilde{\gamma}_\alpha(\Sigma r_i g_i = \Sigma\gamma_\alpha(r_i)g_i$.

It follows that $\Sigma r_i g_i \in \text{Ker } \tilde{\gamma}_\alpha$ if and only if all $r_i \in I_\alpha$. This implies $\Sigma r_i g_i \in I_\alpha(G)$. Then $\text{Ker } \tilde{\gamma}_\alpha = I_\alpha(G)$, whence $R_\alpha(G) \simeq R(G)/I_\alpha(G)$ and $\cap I_\alpha(G) = (\cap I_\alpha)(G) = 0$. Thus the lemma is proved. □

Without doubt the following result is of great value in associative algebras.

LEMMA 1.3: If S is a central simple algebra over F and A is any algebra over F with the same 1, then $G(S \otimes_F A) = S_F \otimes G(A)$.

PROOF: First show that every ideal M of $S \otimes_F A$ is of the form $S \otimes_F K$, where $K \lhd A$. In fact, for each $x \in M$, x is of the form $x = \Sigma s_i \otimes a_i$, $s_i \in S$, $a_i \in A$. However, $S_L \otimes 1 = (S \otimes L)_L$, $(S_R \otimes 1) = (S \otimes 1)_R$. Now let $L(S)$ denote the ring of linear transformations of S regarded as a vector space over F. Obviously, $S \otimes 1 \in L(S \otimes_F A)$ for all $s \in S$. Suppose that $s_1, s_2, \ldots, s_n$ are linearly independent over F. By the density of $L(S)$, for each $s \in S$, there exists $m \in L(S)$, such that $s_1 m = s$, $s_i m = 0$, $i = 2,3,\ldots,n$. Whence $x(m \otimes 1) = s \otimes a_1 \in M$.

Now let $K = \{a \in A \mid \text{for all } s \in S \text{ there is } s \otimes a \in M\}$. As indicated above, $a_1 \in K$; similarly, $a_i \in K$, $i = 2,\ldots,n$ too. Thus $M \subset S \otimes K$, and clearly $M = S \otimes K$. It is easy to verify that $K \lhd A$.

Next if $K \lhd A$, then $S \otimes K \lhd S \otimes A$. Thus the map $K \to S \otimes K$ is a lattice isomorphism of the ideal lattice of A onto that of $S \otimes_F A$. Moreover, $K_1 \subset K_2$ if and only if $S \otimes K_1 \subset S \otimes K_2$ and K is a maximal ideal of A if and only if $S \otimes K$ is a maximal ideal of $S \otimes A$. We further obtain that $S \otimes A/S \otimes K$ is a G-semisimple ring if and only if A/K is a G-semisimple simple ring.

Finally, $G(S \otimes A) = \cap\{M_\alpha \mid \alpha \in \Omega$, M_α is a maximal ideal of $S \otimes A$, $S \otimes A/M_\alpha$ is G-semisimple$\} = \cap\{S \otimes_F K_\alpha \mid \alpha \in \Omega$, K_α is a maximal ideal of A, A/K_α is G-semisimple$\} = S \otimes_F \cap\{K_\alpha \mid K_\alpha$ is a maximal ideal of A, A/K_α is G-semisimple$\}$ $= S \otimes GA$. □

Now go back to the G-semisimplicity of group rings.

LEMMA 1.4: Let R be a simple ring with 1 and G an abelian group. If ch. R = 0 or ch. R = p and $p \notin O(G)$, then R(G) is G-semisimple.

PROOF: Let the field F be the centre of the simple ring R, then R can be regarded as a central simple algebra over F. Since G is an abelian group and F(G) is an algebra over F, thus F(G) is a free algebra with elements of G as free basis. On the other hand, every element $\alpha \in R \otimes F(G)$ can be expressed uniquely in the form: $\alpha = \Sigma\, a_g g$ $(g \in G,\ a_g \in R)$. Suppose $\gamma: \Sigma\, a_g g \to \Sigma a_g \otimes g$; then it can be verified easily that $R(G) \simeq R \otimes_F F(G)$. By Lemma 1.3 and [2], Corollary 7.3.2., we further obtain that $GR(G) \simeq R \otimes_F GF(G) = R \otimes_F JF(G) = 0$. □

THEOREM 1.3: Suppose that R is a G-semisimple ring with 1 and G is an abelian group. If ch. R = n and each prime divisor of n is not included in O(G), then $GR(G) = 0$.

PROOF: Since R is G-semisimple, then R is a subdirect sum of simple rings R_α with 1, that is $R = \sum_{\alpha\in\Omega} s \oplus R$. Thus R_α is a G-semisimple simple ring. By Lemma 1.2, $R(G) \simeq \sum_{\alpha\in\Omega} s \oplus R_\alpha(C)$.

Suppose ch. $R_\alpha = r$. Since R_α is a homomorphic image of R, it can be deduced easily that r is a prime divisor of n and $r \notin O(G)$. Hence $GR_\alpha(G) = 0$ by Lemma 1.4.

Let γ_α be a homomorphism of R(G) onto $R_\alpha(G)$. Since Ker $\gamma_\alpha = I_\alpha(G)$, where I_α is the homomorphic kernel of R onto R_α, then $R_\alpha(G) \simeq R(G)/I_\alpha(G)$. Further $\bigcap_{\alpha\in\Omega} I_\alpha(G) = 0$. As $[GR(G) + I_\alpha(G)]/I_\alpha(G)$ is a G-regular ideal of $R(G)/I_\alpha(G)$ hence $GR(G) \subset I(G)$ for all $\alpha \in \Omega$ and $GR(G) \subset \bigcap_{\alpha\in\Omega} I_\alpha(G) = 0$. □

COROLLARY 1.1: Suppose R is an associative ring with 1 and G is an abelian group. If ch. R = n and for each prime divisor p of n, $p \notin O(G)$, then $GR(G) \subset (GR)(G)$.

PROOF: Let $\bar{R} = R/GR$, then $\bar{R}$ is G-semisimple and ch. $\bar{R}$ is a divisor of ch. R. Whence for any prime p, if p is a divisor of ch. $\bar{R}$, certainly p is a divisor of n and $p \notin O(G)$. It follows that $G\bar{R}(G) = 0$.

Let $\gamma:R \to \bar{R}$ be the natural homomorphism; then γ can be extended to homomorphism $\tilde{\gamma}:R(G) \to \bar{R}(G)$ given by $\tilde{\gamma}(\Sigma a_g g) = \Sigma\, \gamma(a_g)g$.

It is easy to verify that Ker $\tilde{\gamma}$ = $(GR)(G)$; thus $R(G)/(GR)(G) \simeq \bar{R}(G)$ and $R(G)/GR(G)$ is G-semisimple. Since $[(GR(G) + (GR)(G)]/(GR)(G)$ is a G-regular ideal of $R(G)/(GR)(G)$, then $(GR(G) \subset (GR)(G)$. □

2. FINITE GROUPS

We first discuss the control of J-radical, then prove that Farahat's equality holds for the G-radical if G is a finite group. Suppose S is a subring of R with the same 1.

Now let $T = \{x_i \in R \mid i \in \Omega\}$. Then T is called the normal basis of R over S if each $\alpha \in R$ can be expressed uniquely in the form of a finite sum $\alpha = \Sigma\, s_i x_i$, where $s_i \in S$ and there exist automorphisms σ_{x_i} of S such that $\sigma_{x_i}(\beta)x_i = x_i\beta$, $x_i \in T$. In addition, let M be an R-module and W an R-submodule of M. If W regarded as S-module is a direct summand of M implies that W regarded as R-module is a direct summand of M, we say that R is S-projective.

LEMMA 2.1: Suppose R is an associative ring with 1, $H \triangleleft G$ and $[G : H] = n < \infty$. Then:

(i) $(JR(G))^n \subset JR(H).R(G) \subset JR(G)$.

(ii) If n is a unit in R, then R(G) is R(H)-projective.

(iii) If R(G) is R(H)-projective, then

$$JR(H).R(G) = JR(G).$$

PROOF: Let $\mathcal{D} = \{g_1, g_2, \ldots, g_n\}$ be a representative set of cosets of H in G and $\sigma_g:\beta \to g\beta g^{-1}$, for any $\beta \in R(H)$. Then σ_g is an automorphism of R(H) and $\sigma_g(\beta)g = g\beta$. Hence $\mathcal{D}$ is a normal basis of R(G) over R(H). Evidently both R(G) and R(H) have the same 1. (i) is trivial by [2], Theorem 7.2.5.

Let M be an R(G)-module and W an R(G)-submodule. If W regarded as R(H)-module is a direct summand of M, that is $M_{R(H)} = W_{R(H)} \oplus V_{R(H)}$, it follows that there exists an R(H)-module projection f of M into W such that $f(w) = w$ and $f(m\alpha) = f(m)\alpha$, for all $w \in W$, $m \in M$, $\alpha \in R(H)$. Now define γ as follows: $\gamma(m) = (1/n)\, \Sigma f(mg_i^{-1})g_i$ for all $m \in M$. It is easy to show that

$\gamma(m)g = \gamma(mg)$, $\gamma(m\beta) = \gamma(m)\beta$ for all $\beta \in R(G)$, whence γ is an $R(G)$-module projection of $R(G)$ onto W. Suppose $\upsilon = \{u \in M \mid \gamma(u) = 0\}$. Then we can deduce easily that υ is an $R(G)$-submodule of M and $M = W \oplus V$. This completes the proof of (ii) (cf. [2], Lemma 7.2.2.). (iii) can be obtained immediately by [2], Theorem 7.2.5. □

THEOREM 2.1: Suppose that R is an associative ring with 1, $H \lhd G$ and G/H is locally finite. If every $n \in O(G/H)$ is a unit in R, then $JR(G) = JR(H).R(G)$.

Particularly, if G is locally finite and every $n \in O(G/H)$ is a unit in R, then $JR(G) = (JR)(G)$.

PROOF: Let $L = \langle H, \text{supp}\ \alpha\rangle$, for any $\alpha \in JR(G)$. Then $\alpha \in R(L)$ and $\alpha \in JR(G) \cap R(L) \subset JR(L)$. Since L/H is a finite group and $|L/H|$ is a unit in R, by (ii) and (iii) of Lemma 2.1, we get $\alpha \in JR(L) = JR(H).R(L) \subset JR(H)R(G)$. It implies that $JR(G) \subset JR(H).R(G)$.

Conversely, since $\alpha R(G)$ is a right ideal of $R(G)$ for any $\alpha \in JR(H)$ and let $G_\beta = \langle H, \text{supp}\ \beta\rangle$ for any $\beta \in R(G)$, then G_β/H is a finite group and $|G_\beta/H|$ is a unit in R. By Lemma 2.1 we know that $\alpha\beta \in JR(H).R(G_\beta) = JR(G_\beta) \subset R(G)$. This implies that $\alpha\beta$ is a quasiregular element in $R(G_\beta)$; obviously it is the case in $R(G)$, too. Hence $\alpha R(G) \subset JR(G)$ and $JR(H).R(G) \subset JR(G)$. □

Therefore, H may control $JR(G)$ under the J-radical. Now let us go back to the G-radical.

COROLLARY 2.1: Suppose that H is a commutative ring with 1 and H is an abelian normal subgroup of G. If G/H is locally finite and every element of $O(G)$ is a unit in R, then $\mathcal{G}R(H).R(G) \subset \mathcal{G}R(G)$.

In fact, if $R(H)$ is commutative, then $\mathcal{G}R(H)R(G) = JR(H)R(G) = JR(G) \subset \mathcal{G}R(G)$ by Theorem 2.1.

COROLLARY 2.2: Suppose that R is a commutative ring with 1 and G is a locally finite group. If any $n \in O(G)$ is a unit in R, then $(\mathcal{G}R)(G) \subset \mathcal{G}R(G)$.

By Corollary 1.1 and Corollary 2.2, we can obtain the following corollary immediately.

COROLLARY 2.3: Suppose that R is a commutative ring with 1, G locally finite abelian group. If any $n \in O(G)$ in a unit in R, then

$$(\mathcal{G}R)(G) = \mathcal{G}(R(G)).$$

It is well-known that every torsion abelian group is a locally finite abelian group.

In the following theorem we give a criterion of R(G) being $\mathcal{G}$-semisimple if G is finite group and Farahat's equality is true for Brown-McCoy radical.

THEOREM 2.2: Suppose that R is an associative ring with 1 and G is a finite group. If ch. R = m and each prime divisor of m is not contained in O(G), then

(i) R(G) is also $\mathcal{G}$-semisimple if R is $\mathcal{G}$-semisimple;

(ii) $\mathcal{G}R(G) = (\mathcal{G}R)(G)$.

PROOF: First show the case that R is a simple ring, ch. R = 0 or ch. R = p and $p \notin O(G)$. Let the field F be the centre of the simple ring R. Then obviously, ch. R = ch. F. By the proof of Lemma 1.4 we know $R(G) \simeq R \otimes_F F(G)$. Moreover, $\mathcal{G}R(G) = R \otimes_F \mathcal{G}F(G)$ by Lemma 1.3. Since the ideals of F(G) regarded as an algebra are identical with those of F(G) regarded as a ring, therefore F(G) satisfies the Artinian condition. This implies that $\mathcal{G}F(G) = JF(G)$. By [2], Theorem 2.4.2 (Maschke's theorem), we obtain $JF(G) = 0$. It follows that $\mathcal{G}R(G) = R \otimes_F \mathcal{G}F(G) = 0$.

Now let us turn to the general case (i), since R is $\mathcal{G}$-semisimple, then R is a subdirect sum of simple R_α, with 1 which is denoted by

$$R = \sum_{\alpha\in\Omega} s \oplus R_\alpha.$$

By Lemma 1.2,

$$R(G) = \sum_{\alpha\in\Omega} s \oplus R(G).$$

Let ch. $R_\alpha = 0$ or ch. $R_\alpha = p_\alpha$ (prime) (hence p_α must divide m, and $p_\alpha \notin O(G)$). Thus $GR_\alpha(G) = 0$. By a similar proof of Theorem 1.3 we obtain $GR(G) = 0$.

The proof of $GR(G) \subset (GR)(G)$ in (ii) is quite similar to that of Corollary 1.1, on the other hand, becuase G is finite group, it is clear that R(G) is a noraml extension of R and $GR = GR(G) \cap R$ (cf. [10]), hence $(GR)G \subset GR(G)$, (ii) is true. □

3. RESIDUALLY FINITE GROUPS

Residually finite group means that for any $e \neq g \in G$ there exists a normal subgroup N_g of G such that $g \notin N_g$ and G/N_g is a finite group.

LEMMA 3.1: A group G is a residually finite group if and only if there exists a family $\{G_\lambda \mid \lambda \in \Omega\}$ of normal subgroup of G such that

(i) $\bigcap_{\lambda\in\Omega} G_\lambda = \{e\}$,

(ii) G/G_λ is finite,

(iii) for any $\sigma, \gamma \in \Omega$, there exists a $\lambda \in \Omega$ such that $G_\lambda \subset G_\sigma \cap G_\gamma$ (cf. [2]).

Polycyclic groups are residually finite, and both finitely generated linear groups and free groups are residually finite. Therefore it is of value to study residually finite groups.

Let H be a subgroup of G and wH be the ideal generated by $\{1 - g \mid g \in H\}$ in R(G). Then $wH \lhd R(G)$ if and only if $H \lhd G$ [1]. Let $T = \{G_\alpha \lhd G \mid \alpha \in \Omega\}$ and $\Pi = \{|G/G_\alpha|, G_\alpha \in T\}$. If T satisfies (i)-(iii) of Lemma 3.1 above, T is called a residually finite family, and Π is called a residually finite exponential set. Let $\gamma_\alpha : R(G) \to R(G/G_\alpha)$ be homomorphisms given by $\gamma_\alpha(\Sigma a_g g) = \Sigma a_g gG_\alpha$.

LEMMA 3.2: Let R be an associative ring with 1, G be a residually finite group and $T = \{G_\alpha \lhd G \mid \alpha \in \Omega\}$ be a residually finite family of G, then $\bigcap_{\alpha\in\Omega} wG_\alpha = 0$.

PROOF: It is easy to show that Ker $\gamma_\alpha = wG_\alpha$ by [1]. Take an arbitrary $x = \Sigma a_i g_i \in \cap wG$, where $a_i \in R$, $g_i \in G$, such that $g_i \neq g_j$ for $i \neq j$.

Obviously, there is $\gamma_\alpha(x) = \Sigma a_i(g_iG) = 0$ for all $\alpha \in \Omega$.

On the other hand, if $g_ig_j^{-1} \in G_\alpha$ for all $\alpha \in \Omega$, then $g_ig_j^{-1} \in \bigcap_{\alpha\in\Omega} G_\alpha = \{e\}$, which implies that $g_i = g_j$. This is a contradiction. Thus if $i \neq j$, there exists $\lambda_{(i,j)}$ such that $g_ig_j^{-1} \in G_{\lambda(i,j)}$. By Lemma 3.1, there exists μ such that $G_\mu \subset \bigcap_{i\neq j} G_{(i,j)}$ and $g_ig_j^{-1} \notin G_\mu$, for all $i,j \in \Omega$, $i \neq j$. Hence $g_1G, g_2G,\ldots,g_nG_n$ are different from each other. Since

$$\gamma_\mu(x) = a_1g_1G + a_2g_2G + \ldots + a_ng_nG_\mu = 0,$$

which implies $a_i = 0$ for $i = 1, 2,\ldots,n$ and $\bigcap_{\alpha\in\Omega} wG_\alpha = 0$. □

THEOREM 3.1: Let R be an associative ring with 1, G be a residually finite group and be the residual finite exponential set. If ch. R = m, for each $k \in \Pi$, any divisor p of m does not divide k, then R(G) is G-semisimple if R is G-semisimple.

PROOF: Let $T = \{G_\alpha \triangleleft G \mid \alpha \in \Omega\}$ be the residually finite family of G. Since G/G_α ($\alpha \in \Omega$) is a finite group, and for each $k \in \Pi$, any divisor p of m does not divide k, thus $GR(G/G_\alpha) = 0$ by Theorem 2.2.

On the other hand, let γ_α be the natural homomorphism extension of R(G) onto $R(G/G_\alpha)$. Then $R(G)/wG_\alpha \simeq R(G/G_\alpha)$ for all $\alpha \in \Omega$. This implies that $R(G)/wG_\alpha$ is G-semisimple. Since $(GR(G) + wG_\alpha)/wG_\alpha = 0$, thus we can deduce easily that $GR(G) \subset wG_\alpha$ for all $\alpha \in \Omega$. It follows that $GR(G) \subset \bigcap_{\alpha\in\Omega} wG_\alpha = 0$. □

COROLLARY 3.1: Let R be an associative ring with 1. ch. R = m and G a residually finite group. Suppose that Π is the residually finite exponential set of G and any element of Π is not divisible by p where p is an arbitrary divisor of m, then $GR(G) \subset (GR)(G)$.

PROOF: Since $R = \bar{R}/R$ is G-semisimple and ch. $\bar{R}$ may divide ch. R, hence p divides ch. R. This implies that p does not divide any elements in Π. Since $R(G)/(GR)(G) \simeq (R/GR)(G)$ and the latter is G-semisimple, it is easy to prove that $GR(G) \subset (GR)(G)$. □

REFERENCES

[1] Connell, I.G., On the group ring, Canad. J. Math. 15 (1963), 650-685.

[2] Passman, D.S., The Algebraic Structure of Group Rings, Wiley, New York, 1978.

[3] Zhu Yuan-sen, On semi-simplicity of group rings over ring, J. Math. Res. Expos., China, 2, No. 4 (1982), 31-36.

[4] Passman, D.S., Radical ideals in group rings of locally finite groups, J. Algebra 33 (1975), 472-497.

[5] Wallace, D.A.R., Some applications of subnormality in groups in the study of group algebras, Math. Z. 108 (1968), 53-62.

[6] Motose, K., On group rings over semi-primary rings II, J. Fac. Sci. Shinshu Univ. 6 (1971), 97-99.

[7] Coleman, D.B., On group rings, Canad. J. Math. 22 (1970), 249-254.

[8] Groenewald, N.J., On the Brown-McCoy radical of group rings, Publ. Math. Beograd (NS) 27 (41) (1980), 57-59.

[9] Knott, R.P., The Brown-McCoy radical of certain group algebras, J. Lond. Math. Soc. (2) 6 (1973), 617-625.

[10] Nastasescu, C., and Van Oystaeyen, F., Graded Ring Theory, North-Holland 1982.

Zhu Yuan-sen
Department of Mathematics
Hebei Teachers' University
Shi-Jia-Zhuang
Prov. Hebei
China

ZHU YUAN-SEN

H.K. Farahat's problem concerning the radicals of group rings

In this paper we use R(G) to denote the group ring of a group G over a ring R. There have not been any good results about how to describe the radicals of R(G). However, for the group algebra F(G), where F is a field, it has been shown in [4] and [5], for some special groups, that JF(G) may be controlled by some subgroup H of G, i.e. JF(G) = JF(H).F(G). Furthermore, H.K. Farahat raised the problem, "When is JR(G) = (JR)(G)?" Obviously it is very valuable for describing the radical of a group ring, and by means of this result the semisimplicity of RG can be changed into semisimplicity of the base ring. In [6] and [7] it has been proved that Farahat's equality holds for the J-radical when G is locally finite, R is semiprimary and commutative respectively. In [3], the result has been generalized to general associative rings, and for the Brown-McCoy radical the Farahat equality is also valid when G is finite or locally finite abelian.

In this paper, we mainly prove that the Farahat equality holds for the Baer, Levitzki and Köthe radicals when R is commutative, for any group G. At the same time it is shown that the Farahat equality also holds when R is artinian and G is any group such that F(G) is J-semisimple and for any R when G is locally finite.

1. THE BAER RADICAL

Let $\mathcal{B}R$ denote the Baer radical of a ring R.

LEMMA 1.1: If R is a commutative ring with identity and G any group, then $(\mathcal{B}R)(G) \subset \mathcal{B}R(G)$ and $(\mathcal{B}R)(G)$ is a $\mathcal{B}$-radical ring.

PROOF: By [1], Proposition 9, we know that $\mathcal{B}R = R \cap \mathcal{B}R(G)$ which implies $\mathcal{B}R \subset \mathcal{B}R(G)$. Take an arbitrary element $r = \sum_{i=1}^{m} a_i g_i \in (\mathcal{B}R)(G)$, where $a_i \in \mathcal{B}R$ and $g_i \in G$. Let K be the ideal in R generated by the coefficients $a_1, a_2, \ldots, a_m$ of r. Since $\mathcal{B}R$ is locally nilpotent and commutative, K is nilpotent. Now consider an M-sequence in R(G) starting with r:

$$r_1 = r,\ r_2 = r_1 t_1 r_1,\ \ldots,\ r_j = r_{j-1} t_{j-1} r_{j-1},\ \ldots \qquad (*).$$

Since the coefficients of r_j can be regarded as products of j elements in K, there exists an n such that $r_n = 0$. Thus r is strongly nilpotent so $r \in \mathcal{B}R(G)$ and $(\mathcal{B}R)(G) \subset \mathcal{B}R(G)$.

Since the $\mathcal{B}$-radical class is hereditary, therefore

$$\mathcal{B}((\mathcal{B}R)(G)) = (\mathcal{B}R)(G) \cap \mathcal{B}R(G) = (\mathcal{B}R)(G)$$

whence $\mathcal{B}R(G)$ is a $\mathcal{B}$-radical ring. □

Let $\nu(G)$ be the set of orders of normal subgroups of G.

THEOREM 1.1: Let R be commutative with 1, G a group such that every element of $\nu(G)$ is a unit in R. Then $(\mathcal{B}R)(G) = \mathcal{B}R(G)$.

PROOF: Suppose γ is the natural homomorphism from R onto $R/\mathcal{B}R$. We extend γ to a homomorphism $\bar{\gamma}$ from R(G) onto $(R/\mathcal{B}R)(G)$ such that $\bar{\gamma}(\Sigma\, a_i g_i) = \Sigma\gamma(a_i)g_i$, where $a_i \in R$, $g_i \in G$. Obviously Ker $\bar{\gamma} = (\mathcal{B}R)(G) \lhd R(G)$ and $R(G)/(\mathcal{B}R)G \simeq (R/\mathcal{B}R)(G)$. By [1], Theorem 5, we know that $(R/\mathcal{B}R)(G)$ is $\mathcal{B}$-semisimple. Hence $R(G)/(\mathcal{B}R)(G)$ is also $\mathcal{B}$-semisimple. Therefore $\mathcal{B}R(G) \subset (\mathcal{B}R)(G)$. By Lemma 1.1 the theorem is proved. □

COROLLARY 1.1: Under the conditions of Theorem 1.1, R(G) is $\mathcal{B}$-semisimple if and only if R is $\mathcal{B}$-semisimple.

2. THE LEVITZKI RADICAL

Let $\mathcal{L}R$ be the Levitzki radical of R.

LEMMA 2.1: Suppose K is a locally nilpotent ideal of R and G is any group. Then K(G) is a locally nilpotent ideal of R(G).

PROOF: Certainly K(G) is an ideal of R(G). Take any elements $\alpha_1, \alpha_2, \ldots, \alpha_m \in K(G)$ (finitely many), let $\bar{N} = \langle \alpha_1, \alpha_2, \ldots, \alpha_m \rangle$ be the subring they generate and N the subring of R generated by the coefficients of $\alpha_1, \alpha_2, \ldots, \alpha_m$. It s not difficult to prove that $N \subset K$ and N is thus a nilpotent subring of R. et n be the nilpotent exponent of N. Take any n elements $\beta_1, \beta_2, \dots, \beta_n$ in . Then $\beta_1\beta_2 \dots \beta_n = \Sigma\, a_{i_1} a_{i_2} \dots a_{i_n} \cdot g_{i_1} g_{i_2} \dots g_{i_n}$, where a_{i_1}, $a_{i_2}, \dots, a_{i_n} \in N$. Hence $\beta_1\beta_2 \dots \beta_2 \dots \beta_n = 0$, i.e. $\bar{N}^n = 0$. Therefore K(G) s locally nilpotent. □

LEMMA 2.2: If R has an identity, then for any group G, $(LR)(G) \subset LR(G)$ and $(LR)(G)$ is an L-radical ring.

THEOREM 2.1: Let R be a commutative ring with 1, G any group. If ch. R = 0 or ch. R = n and there is no divisor of n in $\nu(G)$, then R is L-semisimple if and only if R(G) is L-semisimple.

PROOF: Since $LR = 0$, R has no nonzero nilpotent element. From [8], Theorems II, III, we can obtain that R(G) has no nilpotent ideal. Hence $LR(G) = 0$.

Conversely, suppose $LR(G) = 0$. If $LR \neq 0$, then $(LR)(G) \neq 0$. By Lemma 2.2, $0 \neq (LR)(G) \subset LR(G)$. This contradicts $LR(G) = 0$. Therefore R is L-semisimple. □

THEOREM 2.2: Let R be a commutative with 1, G any group. Suppose every element of $\nu(G)$ is invertible in R if ch. R = 0 or that no divisor of ch. R is in $\nu(G)$ if ch. R = n. Then $(LR)(G) = LR(G))$.

PROOF: Suppose $LR = R$. By Lemma 2.1, we know that R(G) is a locally nilpotent ring so obviously we have $LR(G) = R(G) = (LR)(G)$.

Suppose $LR \neq R$. We extend the natural homomorphism of R onto $\bar{R} = R/LR$ to a homomorphism $\bar{\gamma}$ of R(G) onto $\bar{R}(G)$ such that $\bar{\gamma}(\Sigma\, a_i g_i) = \Sigma\, \gamma(a_i) g_i$. It is easy to see that Ker $\bar{\gamma} = (LR)(G)$ and

$$R(G)/(LR)(G) \simeq \bar{R}(G). \qquad (*)$$

Now p is a unit of R if and only if p is a unit of $\bar{R}$.

(i) Suppose ch. R = 0. If ch. $\bar{R} = 0$ then $(\bar{R},+)$ is a torsion-free abelian group and $\bar{R} = R/LR$ has no nilpotent element. Hence $L(\bar{R}(G)) = 0$ by [8], Theorem II.

If ch. $\bar{R} = r \neq 0$, then r is not a unit in $\bar{R}$. Otherwise, r is also a unit in R. Since $r\bar{1} = 0$, $r1 \in LR$, which implies $LR = R$. This contradicts $LR \neq R$. Suppose that p is any divisor of r; let r = pt. If $p \in \nu(G)$, then p is a unit in R whence p is also a unit in $\bar{R}$. Since $0 = r\bar{1} = p\bar{1}.t\bar{1}$ then $t\bar{1} = 0$, which contradicts the fact that r is the characteristic of $\bar{R}$. Hence for any divisor of r, p does not belong to $\nu(G)$. By [8], Theorem III, it follows that $L\bar{R}(G) = 0$.

(ii) Suppose ch. R = n and ch. $(R/LR) = r$. It can be deduced immediately that r is a divisor of n. Therefore if p is a divisor of r, p is also divisor of n, and so p does not belong to $\nu(G)$. We can obtain $L\bar{R}(G) = 0$ by [8], Theorem III. To sum up, under any cases we have $L\bar{R}(G) = 0$, and $R(G)/(LR)(G)$ is L-semisimple by (*). Since $[LR(G) + (LR)(G)]/(LR)(G)$ is a locally nilpotent ideal of $R(G)/(LR)(G)$, then $LR(G) \subset (LR)(G)$. By Lemma 2.2 we have $LR(G) = (LR)(G)$. □

3. THE KÖTHE RADICAL

Let KR denote the Köthe radical of a ring R. For the K-radical we may obtain completely similar results to those for the Levitzki radical, and the proofs are also similar. Therefore we give only the main idea of these proofs.

LEMMA 3.1: Suppose that R is a commutative ring and G any group. Then $(KR)(G) \subset KR(G)$, and $(KR)(G)$ is a K-radical ring.

In fact, when R is a commutative ring we have $KR = LR$. By Lemma 2.2 we may obtain $(KR)(G) \subset LR(G) \subset K(R(G))$.

THEOREM 3.1: Let R be a commutative ring and G any group. If ch. R = 0 or ch. R = n, but p does not belong to $O(G)$ where p is divisor of n, then R is K-semisimple if and only if R(G) is K-semisimple.

(Here $O(G)$ denotes the set of all finite subgroup orders of G.) Actually, if $KR = 0$, by [8], Theorems I, II, it can be shown that R(G) does not contain any nonzero nil ideal. Then $KR(G) = 0$. The converse may be shown by Lemma 3.1.

THEOREM 3.2: Let R be a commutative ring with 1, G any group. If every element of $O(G)$ is a unit in R when ch. $R = 0$ and any prime divisor of n does not belong to $O(G)$ when ch. $R = n$, then $(KR)(G) = K(R)(G)$.

In fact, if $KR = R$, the theorem is obviously true; if $KR \neq R$, let $\bar{R} = R/KR$. Then $R(G)/(KR)(G) \simeq \bar{R}(G)$. Suppose ch. $R = 0$. If ch. $\bar{R} = 0$, it may be obtained that $K\bar{R}(G) = 0$ by [8], Theorem II. If ch. $\bar{R} = r \neq 0$, then r is not a unit in R. For every prime divisor p of r and $r = pt$, if p belongs to $O(G)$, it may be deduced that $t\bar{1} = 0$ and $t < r$, a contradiction. Thus $p \notin O(G)$. By [8], Theorem I, we have $KR(G) = 0$.

Suppose ch. $R = n$, but ch. $\bar{R} = r$. It may be deduced that any prime divisor of r does not belong to $O(G)$. Using [8], Theorem 1, again, it can be shown that $KR(G) = 0$.

Therefore, under any case we have $K\bar{R}(G) = 0$. Similar to Theorem 1.1, we may prove $KR(G) = (KR)(G)$.

In the three sections above we have completely solved the semisimplicity of group algebra $F(G)$ for the Baer radical, Levitzki radical and Köthe radical. As long as any p of $O(G)$ belongs to F, $F(G)$ is semisimple for the three radicals mentioned above. Thus, to study the radicals and semisimplicity of group algebras and group rings, we need only to study their Jacobson radical, Brown-McCoy radical and semisimplicity.

4. THE JACOBSON RADICAL

From the preceding sections we know to that study when the Farahat equality is true, it suffices to discuss when $R(G)$ is also J-semisimple for R which is J-semisimple. Suppose that F is a field, ch. $F = p$ and $p \notin O(G)$. By [1], [2] we know that $F(G)$ is J-semisimple, if G is a locally finite group, or a residually finite group, or a solvable group, or an ordered group, or free group, etc. Let $C = \{G \mid \text{for any field } F, \text{ if ch. } F \notin O(G), \text{ then } JF(G) = 0\}$.

LEMMA 4.1: Suppose that R is a primitive ring and G is a group such that $JF(G) = 0$. Then $R(G)$ is semiprimitive ring.

PROOF: Since R is primitive ring, let field F be the centre of R. Then $JF(G) = 0$. It follows that $F(G)$ is a separable algebra over F by [2], Lemma 3.11. However, if R is regarded as an algebra over F, clearly R is a semisimple algebra over F. By the Bourbaki theorem we obtain that $R \otimes_F F(G)$ is a semiprimitive ring. On the other hand, $F(G)$ is regarded as a free F-module with basis $\{g \mid g \in G\}$. However each element of $R \otimes_F F(G)$ can be written uniquely in the form $\Sigma\, a \otimes g$, where $a \in R$, $g \in G$. This implies that the map $\Sigma\, ag \to \Sigma\, a \otimes g$ $(a \in R, g \in G)$ is a ring isomorphism of $R(G)$ onto $R \otimes_F F(G)$. Hence $R(G)$ is a semiprimitive ring. □

THEOREM 4.1: Let R be a semiprimitive ring, $G \in C$. If ch. R = n and no prime divisor of n belongs to $O(G)$, then $JR(G) = 0$.

PROOF: Let $\{M_\mu \mid \mu \in I\}$ be the set of all maximal ideals of R. Since $\bar{R}_\mu = R/M_\mu$ is a simple ring with 1, then $\bar{R}_\mu$ is a primitive ring. Hence each M_μ is a primitive ideal of R. Since $n\bar{1} = 0$, let $ch.R_\mu = p$. It is clear that p is a divisor of n and $p \notin O(G)$. If F is the centre of $\bar{R}_\mu$, then F is a field and ch. $\bar{R}$ = ch. F, whence $G \in C$ and $F(G)$ must be a semiprimitive ring. By Lemma 4.1, $\bar{R}_\mu(G)$ is a semiprimitive ring.

We extend natural homomorphism $R \to \bar{R}_\mu = R/M_\mu$ to homomorphism $\gamma: R(G) \to \bar{R}_\mu(G)$ such that $\gamma(\alpha) = \gamma(\Sigma\, a_g g) = \Sigma(a_g + M_\mu)g$. It is not difficult to verify that $\operatorname{Ker} \gamma = M_\mu(G)$. Since

$$R(G)/M_\mu(G) \simeq \bar{R}_\mu(G)$$

then $R(G)/M_\mu(G)$ is a semiprimitive ring and $JR(G) \subset M_\mu(G)$. It follows that $JR(G) \subset \bigcap_{\mu\in I} M_\mu(G) = (\bigcap_{\mu\in I} M_\mu)(G) = (JR)(G)$, whence $R(G)$ is a semiprimitive ring. □

COROLLARY 4.1: Suppose that R is an associative ring with 1, $G \in C$, ch. R = n and for any prime divisor of n we have $p \notin O(G)$. Then $JR(G) \subset (JR)(G)$.

PROOF: Let $\bar{R} = R/JR$. We extend the natural homomorphism $R \to \bar{R}$ to a homomorphism of $R(G)$ onto $\bar{R}(G)$ whose kernel is $(JR)(G)$, then $R(G)/(JR)(G) \simeq \bar{R}(G)$.

Since $n\bar{1} = 0$, if ch. $\bar{R} = r$, then r is a divisor of n, whence $p \notin O(G)$ for every prime divisor p of n. By Theorem 1.1, $R(G)/(JR)(G)$ is semiprimitive. Therefore $JR(G) \subset (JR)(G)$. □

From this we know that the Farahat equality has been changed into the condition under which $(JR)(G) \subset JR(G)$.

LEMMA 4.2:

(i) Suppose that R is an artinian ring with 1, G any group or

(ii) R is an associative ring, and G is locally finite group. Then $(JR)(G) \subset JR(G)$.

PROOF: In the two cases above we have $JR = R \cap JR(G)$ by [1], Proposition 9. Then $JR \subset JR(G)$. Moreover, for every $r \in JR$, $g \in G$, we have $rg \in JR(G)$, from which it follows that $\Sigma r_i g_i \in JR(G)$ where $r_i \in JR$, $g_i \in G$. Hence $(JR)(G) \subset JR(G)$. □

To sum up, we obtain the following results.

THEOREM 4.2: Suppose that R is an artinian ring with 1, $G \in C$, and no prime divisor of n belongs to O(G) if ch. R = n. Then $JR(G) = (JR)(G)$.

THEOREM 4.3: Suppose that R is associative ring with 1, G locally finite, and no prime divisor of n belongs to O(G) if ch. R = n. Then $JR(G) = (JR)(G)$.

REFERENCES

[1] Connell, I.G., On the group ring, Canad. J. Math. 15 (1963), 650-685.

[2] Passman, D.S., Structure of Group Rings, Wiley, New York, 1978.

[3] Zhu Yuan-sen, Brown-McCoy semisimplicity of group rings, this volume, previous chapter.

[4] Passman, D.S., Radical ideals in group rings, J. Algebra 33 (1975), 472-497.

[5] Wallace, D.A.R., Some applications of subnormality in groups in the study of group algebra, Math. Z. 108 (1968), 53-62.

[6] Motose, K., On group rings over semi-primar rings II, J. Fac. Sci. Shinshu Univ. 6 (1971), 97-99.

[7] Coleman, D.B., On group rings, Canad. J. Math. 22 (1975), 249-254.

[8] Passman, D.S., Nil ideals in group rings, Michigan Math. J. 9 (1962), 375-384.

[9] Divinsky, N.J., Rings and Radicals, University of Toronto Press, Toronto, 1965.

Zhu Yuan-sen
Department of Mathematics
Hebei Teachers' University
Shi-Jia-Zhuang
Prov. Hebei
China

Problems

G.L. BOOTH

Let M be a Γ-ring with left and right operator rings R and L. Then $N(L)^+ \subseteq N(M)$ and $N(R)^* \subseteq N(M)$, where N is the nil radical class. If the Köthe conjecture is correct, then $N(L)^+ = N(M) = N(R)^*$. Is this pair of equalities *equivalent* to the Köthe conjecture?

B.J. GARDNER

1. Does there exist a special radical class R of associative rings, with semisimple class S, such that

$$R \neq U(\cap \{M : M \text{ is a special class of prime rings and } R = U(M)\})?$$

(Here U( ) denotes the upper radical.)

2. Do there exist special classes M_1, M_2 of prime rings for which

$$U(M_1) = U(M_2) \neq U(M_1 \cap M_2)?$$

3. Do there exist disjoint special classes M_1, M_2 of prime rings with $U(M_1) = U(M_2)$?

RÜDIGER GÖBEL

Is the class of hypercotorsion abelian groups (i.e. the lower radical class defined by the class of cotorsion groups) closed under arbitrary direct products? (It is closed under *countable* direct products.)

JONATHAN S. GOLAN

1. Consider the category of all pairs (R,τ) where R is a ring with identity and τ is a torsion theory on R-mod. A morphism from (R,τ) to (S,σ) is a ring homomorphism $R \to S$ such that a left S-module N is σ-torsion if

and only if ${}_R N$ is τ-torsion. What are the radicals in this context?

2. If S is a semiring which is also a complete lattice (with addition in $S = \wedge$ in the lattice) what is the "best" definition of the Jacobson radical of S?

MELVIN HENRIKSEN

Let R be a ring with identiy 1. If 1 is the only regular element (= nonzero-divisor) of R, then R is called a UR-*ring*. A subring of a UR-ring that contains 1 is called an SUR-*ring*. Each SUR-ring R has the following properties:

$$\left.\begin{array}{l} R \text{ is reduced (i.e. } a^2 = 0 \Rightarrow a = 0). \\ 1 + 1 = 0. \\ \text{Every element of } R\setminus\{1\} \text{ is contained in a proper completely prime} \\ \text{ideal of } R \text{ (i.e. an ideal } P \neq R \text{ for which } R/P \text{ has no zerodivisors).} \end{array}\right\} (*)$$

1. Is there a noncommutative UR-ring?

2. Is the centre of a UR-ring always a UR-ring? (Yes if R satisfies a polynomial identity over its centre.)

3. Is every ring with identity which satisfies (*) a SUR ring? (Yes if R is commutative.)

4. Is the free $\mathbb{Z}_2$-algebra (with identity) on two generators a SUR-ring?

PAUL E. JAMBOR

Let $G = [0,\infty)$ be the monoid of nonnegative real numbers with respect to addition and let $S = K[G]/([1,\infty))$ for a field K, where $([1,\infty))$ is the ideal generated by $[1,\infty)$. (Thus S is the "Zassenhaus algebra with identity" over K.) Then

(i) S is a valuation ring (linearly ordered lattice of ideals) and every ideal is of the form xS or xJ(S), $x \in [0,1)$;

(ii) $J(S))^2 = J(S)$ and every proper ideal of J(S) is nilpotent;

(iii) $J(S)$ is not T-nilpotent and

(iv) S is not self-injective.

Does every nonzero S-module possess a nonzero indecomposable direct summand?

BARBARA L. OSOFSKY

1. Let R be a ring with identity. If for all idempotents $e,f \in R$, $eR + fR$ is projective, is $eR \cap fR$ a direct summand of R?

2. Under ($V = L$) all Whitehead groups are free. Under ($MA + \neg CH$) there exists $\aleph_1$-generated nonfree Whitehead groups. What is a nice set-theoretic *equivalent* of "all Whitehead groups (or all of cardinality $\aleph_1$) are free"?

A.D. SANDS

Let $\mathcal{R}$ be a radical class (of associative rings) with semisimple class $\mathcal{S}$. Consider the following condition.

$$R \in \mathcal{R} \text{ implies } R \text{ has no nonzero left ideals in } \mathcal{S}. \qquad (C)$$

1. Does right-hereditary + (C) imply left-hereditary?

2. Does left-hereditary imply right-hereditary for subradicals of the class of hereditarily idempotent rings?

S. VELDSMAN

1. Does Sands' theorem (a nonempty class is semisimple if and only if it is regular, coinductive and extension-closed) hold for groups?

2. Is there a universal class (of whatever "objects") in which all semisimple classes are hereditary (or in which semisimple = coradical) but not all radical classes have the ADS-property?

3. Is there a universal class in which Sands' theorem holds for semisimple classes but not all semisimple classes are hereditary?

R. WIEGANDT

Let V be a universal class of not necessarily associative algebras over a commutative ring R with 1. Further, let us consider the following conditions which V may satisfy:

(B) To every $C \triangleleft B \triangleleft A \in V$ there exist natural numbers m,n,t such that

$$aC^{(m)} + C^{(m)}a \subseteq C,$$

$$(\langle aC + Ca + C\rangle_B)^{(n)} \subseteq C$$

and

$$\langle C^{(t)}\rangle_B \subseteq C^2,$$

where $C^{(m)} = C^{(m-1)}C^{(m-1)}$ and $C^{(0)} = C$, and $\langle X\rangle_B$ denotes the ideal of B generated by X.

(TO) If C is a distinguished ideal of B and $B \triangleleft A \in V$, then also $C \triangleleft A$. (C is a *distinguished* ideal of B if $C \triangleleft B$ and B does not contain ideals E and D such that $E \subsetneqq C \subsetneqq D, C/E \simeq D/C$ and in B/C the relation $\triangleleft$ is transitive.)

Condition (B) is due to Beidar, and condition (TO) is a modification by Puczyłowski and Veldsman of a condition introduced by Terlikowska-Osłowska. These conditions are very effective in proving general radical theoretic results in V and each ensures that every radical class has the ADS-property, that Sands' theorem holds and that the lower radical construction terminates at the ω_0 stage.

Does either of (B), (TO) imply the other?